Fuel Cell Systems Explained

Fuel Cell Systems Explained

James Larminie
Oxford Brookes University, UK

Andrew Dicks
BG Technology, Loughborough, UK

JOHN WILEY & SONS, LTD
Chichester • Weinheim • New York • Brisbane • Singapore • Toronto

Other Wiley Editorial Offices

John Wiley & Sons, Inc., 605 Third Avenue,
New York, NY 10158-0012, USA

Wiley-VCH Verlag GmbH
Pappelallee 3, D-69469 Weinheim, Germany

Jacaranda Wiley Ltd, 33 Park Road, Milton,
Queensland 4064, Australia

John Wiley & Sons (Canada) Ltd, 22 Worcester Road
Rexdale, Ontario, M9W 1L1, Canada

John Wiley & Sons (Asia) Pte Ltd, 2 Clementi Loop #02-01,
Jin Xing Distripark, Singapore 129809

British Library Cataloguing in Publication Data

A catalogue record for this book is available from the British Library

ISBN 0 471 49026 1

Produced from camera-ready copy supplied by the authors
Printed and bound in Great Britain by Bookcraft (Bath) Ltd
This book is printed on acid-free paper responsibly manufactured from sustainable
forestry, in which at least two trees are planted for each one used for paper production.

Contents

Appendices

Foreword

By Dr Gary Acres OBE, formerly Director of Research, Johnson Matthey Plc.

A significant time generally elapses before any new technological development is fully exploited. The fuel cell, first demonstrated by Sir William Grove in 1839, has taken longer than most, despite the promise of clean and efficient power generation.

Following Bacon's pioneering work in the 1950s fuel cells were successfully developed for the American manned space programme. This success, together with a policy to commercialise space technology, led to substantial development programmes in America and Japan in the 1970 and 80s, and more recently in Europe. Despite these efforts, which resulted in considerable technical progress, fuel cell systems were seen to be "always five years away from commercial exploitation".

During the last few years of the 20th century much changed to stimulate new and expanding interest in fuel cell technology. Environmental concerns about global warming and the need to reduce CO_2 emissions provided the stimulus to seek ways of improving energy conversion efficiency. The motor vehicle industry, as well as seeking higher fuel efficiencies, is also required to pursue technologies capable of eliminating emissions: the ultimate goal being the zero emission car. The utility industries, following the impact of privatisation and deregulation, are seeking ways to increase their competitive position while at the same time contributing to reduced environmental emissions.

As these developments have occurred, interest in fuel cell technology has expanded. Increased numbers of people from disciplines ranging from chemistry through engineering to strategic analysis, not familiar with fuel cell technology, have needed to become involved. The need by such people for a single, comprehensive and up to date exposition of the technology and its applications has become apparent, and is amply provided for by this book.

While the fuel cell itself is the key component and an understanding of its features is essential, a practical fuel cell system requires the integration of the stack with fuel processing, heat exchange, power conditioning and control systems. The importance of each of these components and their integration is rightly emphasised in sufficient detail for the chemical and engineering disciplines to understand the system requirements of this novel technology.

Fuel cell technology has largely been the preserve of a limited group consisting primarily of electro and catalyst chemists and chemical engineers. There is a need to develop more people with a knowledge of fuel cell technology. The lack of a comprehensive review of fuel cells and their applications has been a limiting factor in the inclusion of this subject in academic undergraduate and graduate student science and engineering courses. This book, providing as it does a review of the fundamental aspects of the technology, as well as its applications, forms an ideal basis for bringing fuel cells into appropriate courses and postgraduate activities.

The first three chapters describe the operating features of a fuel cell and the underlying thermodynamics and physical factors that determine their performance. A good understanding of these factors is essential to an appreciation of the benefits of fuel cell systems and their operating characteristics compared with conventional combustion based technology. A feature of fuel cell technology is that it gives rise to a range of five main types of system, each with its own operating parameters and applications. These are described in Chapters 4 to 6.

The preferred fuel for a fuel cell is hydrogen. While there are applications where hydrogen can be used directly, such as in space vehicles and local transport, in the foreseeable future, for other stationary and mobile applications, the choice of fuel and its conversion into hydrogen rich gas are essential features of practical systems. The range of fuels and their processing for use in fuel cell systems are described in Chapter 7. Chapters 8 and 9 describe the mechanical and electrical components that make up the complete fuel cell plant for both stationary and mobile applications.

This book offers those new to fuel cells a comprehensive, clear exposition and review to further their understanding, and also provides those familiar with the subject a convenient reference. I hope it will also contribute to a wider knowledge about, and a critical appreciation of, fuel cell systems, and thus to the widest possible application of an exciting 21st century technology that could do much to move our use of energy onto a more sustainable basis.

Gary Acres

February 2000

Acknowledgements

The point will frequently be made in this book that fuels cells are highly interdisciplinary, involving many aspects of science and engineering. This is reflected in the number and diversity of companies that have helped with advice, information and pictures in connection with this project. The authors would like to put on record their thanks to the following companies or organisations who have made this book possible:-

        Advanced Power Sources Ltd, UK
        Alstom Ballard GmbH,
        Armstrong International Inc, USA
        Ballard Power Systems Inc, Canada
        BG Technology Ltd, UK
        DaimlerChrysler Corporation
        DCH Technology Inc, USA
        Eaton Corporation, USA
        Epyx, USA
        GfE Metalle und Materialien GmbH, Germany
        International Fuel Cells, USA
        Johnson Matthey Plc, UK
        Hamburgische Electricitäts-Werke AG, Germany
        Lion Laboratories Ltd., UK
        MTU Friedrichshafen GmbH, Germany
        ONSI Corporation, USA
        Paul Scherrer Institute, Switzerland
        Siemens Westinghouse Power Corporation, USA
        Sulzer Hexis AG, Switzerland
        SR Drives Ltd, UK
        Svenska Rotor Maskiner AB, Sweden
        W.L.Gore and Associates Inc, USA
        Zytek Group Ltd, UK

In addition, a number of people have helped with advice and comments to the text. In particular we would like to thank Felix Büchi of the Paul Scherrer Institute, Richard Stone and Colin Snowdon, both from the University of Oxford, and Tony Hern and Jonathan Bromley of Oxford Brookes University, who have all provided valuable comments and suggestions for different parts of this work. Special thanks is also due to Juliet Barrett who did a good deal of manuscript typing. Finally we are also indebted to family, friends and colleagues who have helped us in many ways, and put up with us while we devoted time and energy to this project.

James Larminie, Oxford Brookes University, Oxford

Andrew Dicks, BG Technology, Loughborough

Abbreviations

AC	Alternating current
AES	Air electrode supported
AFC	Alkaline (electrolyte) fuel cell
ASR	Area specific resistance, the resistance of 1 cm^2 of fuel cell. (N.B. total resistance is ASR *divided* by area.)
BLDC	Brushless DC (motor)
BOP	Balance of plant
CFM	Cubic feet per minute
CHP	Combined heat and power
CPO	Catalytic partial oxidation
DC	Direct current
DIR	Direct internal reforming
EC	European Community
EMF	Electro-motive-force
EVD	Electrochemical vapour deposition
FCV	Fuel cell vehicle
GT	Gas turbine
GTO	Gate turn off
HDS	Hydrodesulphurisation
HHV	Higher heating value
IEC	International Electrotechnical Commission
IGBT	Insulated gate bipolar transistor
IIR	Indirect internal reforming
LHV	Lower heating value
LH_2	Liquid (cryogenic) hydrogen
LPG	Liquid petroleum gas
LSGM	Lanthanum, strontium, gallium, and magnesium oxide mixture
MCFC	Molten carbonate (electrolyte) fuel cell
MEA	Membrane electrode assembly
MOSFET	Metal oxide semiconductor field effect transistor
NASA	National Aeronautics and Space Administration
NL	Normal litre, 1 litre at NTP
NTP	Normal temperature and pressure (20 $^{\circ}$C and 1 atm.)
OCV	Open circuit voltage
PAFC	Phosphoric acid (electrolyte) fuel cell
PEM	Proton exchange membrane or polymer electrolyte membrane – different names for the same thing which fortunately have the same abbreviation.
PEMFC	Proton exchange membrane fuel cell or polymer electrolyte membrane fuel cell
PFD	Process flow diagram
PM	Permanent Magnet
ppb	Parts per billion

ppm	Parts per million
PURPA	Public utilities regulatory policies act
PTFE	Polytetrafluoroethylene
PSI	Pounds per square inch
PWM	Pulse width modulation
SCG	Simulated coal gas
SL	Standard litre, 1 litre at STP
SOFC	Solid oxide fuel cell
SPFC	Solid polymer fuel cell (= PEMFC)
SPP	Small power producer
SRM	Switched reluctance motor
SRS	Standard reference state (25 $^{\circ}$C and 1 bar)
STP	Standard temperature and pressure (= SRS)
TEM	Transmission electron microscope
t/ha	Tonnes per hectare annual yield
THT	Tetrahydrothiophene ($C_4H_8O_2S$)
TOU	Time of use
UL	Underwriters' Laboratory
YSZ	Yttria stabilised zirconia

Symbols

a	Coefficient in base 10 logarithm form of Tafel equation also Chemical activity
a_x	Chemical activity of substance x
A	Coefficient in natural logarithm form of Tafel equation also Area
B	Coefficient in equation for mass transport voltage loss
C	Constant in various equations also Capacitance
c_p	Specific heat capacity at constant pressure, in $J.K^{-1}.kg^{-1}$
$\bar{c}_p$	Molar specific heat capacity at constant pressure, in $J.K^{-1}.mol^{-1}$
d	separation of charge layers in a capacitor
e	Magnitude of the charge on one electron, 1.602×10^{-19} Coulombs
E	EMF or open circuit voltage
E^0	EMF at standard temperature and pressure, and with pure reactants
F	Faraday constant, the charge on one mole of electrons, 96 485 Coulombs
G	Gibbs free energy
ΔG^0	Change in Gibbs free energy at standard temperature and pressure, and with pure reactants
ΔG_{T_A}	Change in Gibbs free energy at ambient temperature
$\bar{g}$	Gibbs free energy per mole
$\bar{g}_f$	Gibbs free energy of formation per mole
$(\bar{g}_f)_X$	Gibbs free energy of formation per mole of substance X
H	Enthalpy
$\bar{h}$	Enthalpy per mole
$\bar{h}_f$	Enthalpy of formation per mole
$(\bar{h}_f)_X$	Enthalpy of formation per mole of substance X
I	Current
i	Current density, current per unit area
i_l	Limiting current density
i_o	Exchange current density at an electrode/electrolyte interface
i_{oc}	Exchange current density at the cathode
i_{oa}	Exchange current density at the anode
m	Mass
$\dot{m}$	Mass flow rate
m_x	Mass of substance x

N	Avagadro's number, 6.022×10^{23}
	also Revolutions per second
n	Number of cells in a fuel cell stack
P	Pressure
P_1, P_2	The pressure at different stages in a process
P_X	Partial pressure of gas X
P^0	Standard pressure, 100 kPa
P_{SAT}	Saturated vapour pressure
P_e	Electrical power, only used when context is clear that pressure is not meant.
R	Molar or 'universal' gas constant, 8.314 J.K^{-1}.mol^{-1}
	also Electrical resistance
r	Area specific resistance, resistance of unit area
S	Entropy
$\bar{s}$	Entropy per mole
$(\bar{s})_X$	Entropy per mole of substance X
T	Temperature
T_1, T_2	Temperatures at different stages in a process
T_A	Ambient temperature
T_c	Combustion temperature
t	Time
V	Voltage
V_c	Average voltage of one cell in a stack
V_a	Activation overvoltage
V_r	Ohmic voltage loss
W	Work done
W'	Work done under isentropic conditions
$\dot{W}$	Power
z	Number of electrons transferred in a reaction
α	Charge transfer coefficient
Δ	Change in …
ε	Electrical permitivity
γ	Ratio of the specific heat capacities of a gas
η	Efficiency
η_c	Isentropic efficiency (or compressor or turbine)
ϕ	Relative humidity
λ	Stoichiometric ratio
ω	Humidity ratio
μ_f	Fuel utilisation

1

Introduction

1.1   Hydrogen Fuel Cells - Basic Principles

The basic operation of the hydrogen fuel cell is extremely simple. The first demonstration of a fuel cell was by lawyer-cum-inventor William Grove in 1839, using an experiment along the lines of that shown in Figures 1.1a and 1.1b. In Figure 1.1a water is being electrolysed into hydrogen and oxygen by passing an electric current through it. In Figure 1.1b the power supply has been replaced with an ammeter, and a small current is flowing. The electrolysis is being reversed – the hydrogen and oxygen are recombining, and an electric current is being produced.

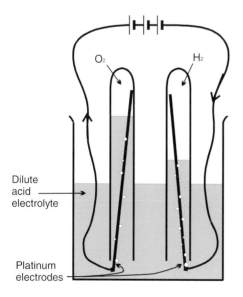
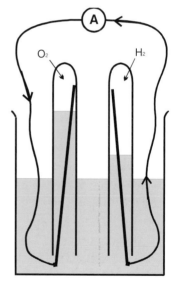

Figure 1.1a   The electrolysis of water. The water is separated into hydrogen and oxygen by the passage of an electric current.

Figure 1.1b   A small current flows. The oxygen and hydrogen are recombining.

Note, the arrows show the direction of flow of the ***negative electrons***, from – to +.

Another way of looking at the fuel cell is to say that the hydrogen fuel is being "burnt" or combusted in the simple reaction:

$$2H_2 + O_2 \rightarrow 2H_2O \tag{1.1}$$

However, instead of heat energy being liberated, electrical energy is produced.

The experiment of Figures 1.1a and 1.1b makes a reasonable demonstration of the basic principle of the fuel cell, but the currents produced are very small. The main reasons for the small current are:

- the low 'contact area' between the gas, the electrode and the electrolyte – basically just a small ring where the electrode emerges from the electrolyte.
- the large distance between the electrodes - the electrolyte resists the flow of electric current.

To overcome these problems the electrodes are usually made flat, with a thin layer of electrolyte as in Figure 1.2. The structure of the electrode is porous, so that both the electrolyte from one side and the gas from the other can penetrate it. This is to give the maximum possible contact between the electrode, the electrolyte and the gas.

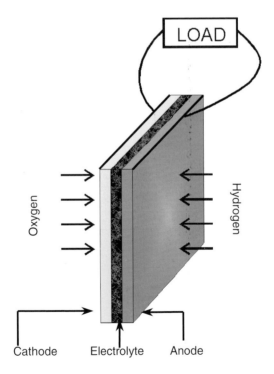

Figure 1.2 Basic cathode-electrolyte-anode construction of a fuel cell

However, to understand how the reaction between hydrogen and oxygen produces an electric current, and where the electrons come from, we need to consider the separate reactions taking place at each electrode. These important details vary for different types of fuel cell, but if we start with a cell based around an acid electrolyte, as used by Grove, we shall start with the simplest and still the most common type.

At the anode of an **acid electrolyte** fuel cell the hydrogen gas ionises, releasing electrons and creating H^+ ions (or protons).

$$2H_2 \rightarrow 4H^+ + 4e^- \qquad [1.2]$$

This reaction releases energy. At the cathode, oxygen reacts with electrons taken from the electrode, and H^+ ions from the electrolyte, to form water.

$$O_2 + 4e^- + 4H^+ \rightarrow 2H_2O \qquad [1.3]$$

Clearly, for both these reactions to proceed continuously, electrons produced at the anode must pass through an electrical circuit to the cathode. Also, H^+ ions must pass through the electrolyte. An acid is a fluid with free H^+ ions, and so serves this purpose very well. Certain polymers can also be made to contain mobile H^+ ions. These materials are called 'proton exchange membranes', as an H^+ ion is also a proton.

Comparing equation 1.2 and 1.3 we can see that two hydrogen molecules will be needed for each oxygen molecule if the system is to be kept in balance. This is shown in Figure 1.3. It should be noted that the electrolyte must only allow H^+ ions to pass through it, and not electrons. Otherwise the electrons would go through the electrolyte, not round the external circuit, and all would be lost.

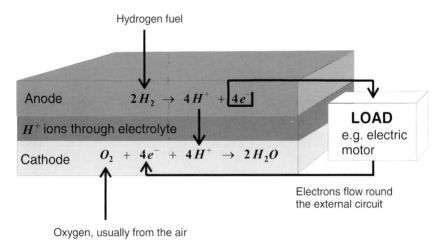

Figure 1.3. Electrode reactions and charge flow for an acid electrolyte fuel cell. Note that although the negative electrons flow from anode to cathode, the "conventional current" flows from cathode to anode

In an **alkaline electrolyte fuel cell** the overall reaction is the same, but the reactions at each electrode are different. In an alkali hydroxyl (OH⁻) ions are available and mobile. At the anode these react with hydrogen, releasing energy and electrons, and producing water.

$$2H_2 + 4OH^- \rightarrow 4H_2O + 4e^-$$ [1.4]

At the cathode oxygen reacts with electrons taken from the electrode, and water in the electrolyte, forming new OH⁻ ions.

$$O_2 + 4e^- + 2H_2O \rightarrow 4OH^-$$ [1.5]

For these reactions to proceed continuously the OH⁻ ions must be able to pass through the electrolyte, and there must be an electrical circuit for the electrons to go from the anode to the cathode. Also, comparing equations 1.4 and 1.5 we see that, as with the acid electrolyte, twice as much hydrogen is needed as oxygen. This is shown in Figure 1.4. Note that although water is consumed at the cathode, it is created twice as fast at the anode.

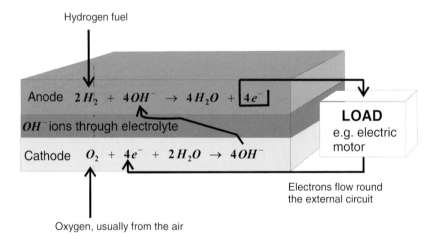

Figure 1.4. Electrode reactions and charge flow for an alkaline electrolyte fuel cell. Electrons flow from anode to cathode, but conventional positive current flows from cathode to anode.

There are many different fuel cell types, with different electrolytes. The details of the anode and cathode reactions are different in each case. However, it is not appropriate to go over every example here. The most important other fuel cell chemistries are covered in Chapter 6 when we consider the solid oxide and molten carbonate fuel cells.

Positive Cathodes and Negative Anodes

Looking at Figures 1.3 and 1.4 the reader will see that the electrons are flowing from the anode to the cathode. The **cathode** is thus the electrically **positive** terminal, since electrons flow from - to +. Many newcomers to fuel cells find this confusing. This is hardly surprising. The Concise Oxford English Dictionary defines cathode as:-

> "**1**. the negative electrode in an electrolyte cell or electron
> valve or tube, **2**. the positive terminal of a primary cell such as
> a battery."

Having two such opposite definitions is bound to cause confusion, but we note that the cathode is the correct name for the positive terminal of **all** primary batteries. It also helps to remember that cations are positive ions, e.g. H^+ is a cation. Anions are negative ions, e.g. OH^- is an anion. It is also true that the **cathode is always the electrode into which electrons flow**, and similarly the anode is always the electrode from which electrons flow. This holds true for electrolysis, cells, valves, forward biased diodes and fuel cells.

A further possible confusion is that while negative electrons flow from minus to plus, the "conventional positive current" flows the other way, from the positive to the negative terminal.

1.2   What Limits the Current?

At the anode hydrogen reacts, releasing energy. However, just because energy is released this does not mean that the reaction proceeds at an unlimited rate. The reaction has the 'classical' energy form shown in Figure 1.5.

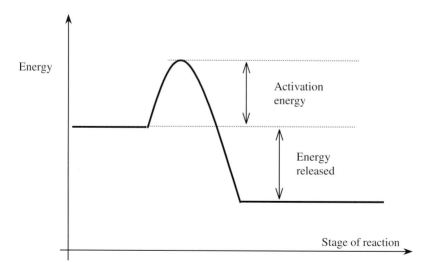

Figure 1.5 Classical energy diagram for a simple exothermic chemical reaction.

Although energy is released, the 'activation energy' must be supplied to get over the 'energy hill'. If the probability of a molecule having enough energy is low, then the reaction will only proceed slowly. Except at very high temperatures, this is indeed the case for fuel cell reactions.

The three main ways of dealing with the slow reaction rates are:
- the use of catalysts,
- raising the temperature,
- increasing the electrode area.

The first two can be applied to any chemical reaction. However, the third is special to fuel cells and is very important. If we take a reaction such as that of equation 1.4, we see that fuel gas and OH^- ions from the electrolyte are needed, as well as the necessary activation energy. Furthermore, this 'coming together' of H_2 fuel and OH^- ions must take place **on the surface of the electrode**, as the electrons produced must be removed. Clearly, the rate at which this happens will be proportional to the area of the electrode. This is very important. Indeed, electrode area is such a vital issue that the performance of a fuel cell design is often quoted in terms of the current *per cm²*.

However, the straightforward area (length × width) is not the only issue. As has already been mentioned, the electrode is made highly porous. This has the effect of greatly increasing the effective surface area. Modern fuel cell electrodes have a microstructure which gives them surface areas that can be hundreds or even thousands of times their straightforward 'length × width' (See Figure 1.6.) The micro-structural design and manufacture of a fuel cell electrode is thus a very important issue for practical fuel cells. As well as these surface area considerations, the electrodes may have to incorporate a catalyst, and endure high temperatures in a corrosive environment. The problems of reaction rates are dealt with in a more quantitative way in Chapter 3.

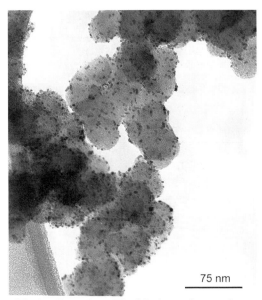

75 nm

Figure 1.6   TEM image of fuel cell catalyst. The black specks are the catalyst particles finely divided over a carbon support. The structure clearly has a large surface area. (Reproduced by kind permission of Johnson Matthey Plc.)

1.3 Connecting Cells in Series – the Bipolar Plate

For reasons explained in Chapters 2 and 3 the voltage of a fuel cell is quite small, about 0.7 volts when drawing a useful current. This means that to produce a useful voltage many cells have to be connected in series. Such a collection of fuel cells in series is known as a 'stack'. The most obvious way to do this is by simply connecting the edge of each anode to the cathode of the next cell all along the line, as in Figure 1.7.

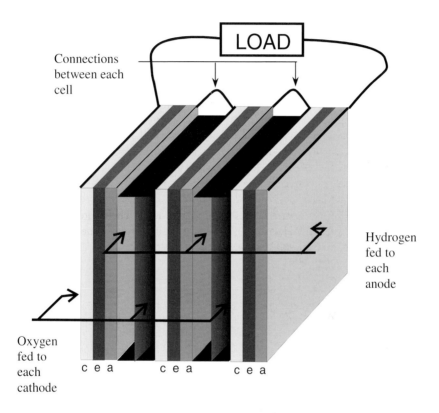

Figure 1.7 Simple edge connection of three cells in series.

The problem with this method is that the electrons have to flow across the face of the electrode to the current collection point at the top edge. The electrodes might be quite good conductors, but if each cell is only operating at about 0.7 volts even a small voltage drop is important. Unless the current flows are very low, and the electrode a particularly good conductor, this method is not used.

A much better method of cell interconnection is to use a 'bipolar plate'. This makes connections all over one cathode and the anode of the next cell (hence 'bipolar'), at the same time the bipolar plate serves as a means of feeding oxygen to the cathode and fuel gas to the anode. Although a good electrical connection must be made between the two electrodes, the two gas supplies must be strictly separated. The general arrangement is shown in Figure 1.8 below.

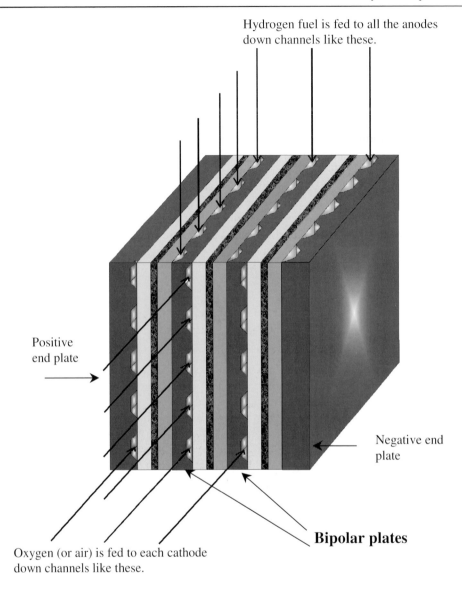

Hydrogen fuel is fed to all the anodes down channels like these.

Positive end plate

Negative end plate

Bipolar plates

Oxygen (or air) is fed to each cathode down channels like these.

Figure 1.8   More efficient way of connecting three cells in series using bipolar plates.

Comparing the bipolar plate arrangement of Figure 1.8 with Figure 1.7 it can easily be seen that the flow of current through the cell is much more efficient. The current is picked up from all over the electrode surface rather than just at the edges. The electrodes are also much more firmly supported, and the whole structure is stronger and more robust. However, the design of the bipolar plate is not simple. If the electrical contact is to be optimised, then the contact points should be as large as possible. However, this would mitigate against good gas flow over the electrodes. If the contact points have to be small, at least they should be frequent. However, this makes the plate more complex, difficult and expensive to manufacture, as well as fragile. Ideally the bipolar plate should be as thin as possible, to minimise electrical resistance, and to make the fuel cells stack small. However, this makes the channels for the gas flow narrow, which means it is more difficult to pump the gas round the cell. This sometimes has to be done at a high rate, especially when using air instead of pure oxygen on the cathode. In the case of low temperature fuel cells, the circulating air has to evaporate and carry away the product water. As well as all this there usually have to be further channels through the bipolar plate to carry a cooling fluid.

The basic form of the bipolar plate is a flat piece with carefully shaped grooves in it for the gas flow. (For this reason they are sometimes also called 'flow field plates'.) The shape of the grooves is usually much more complex than the simple straight channels implied by Figure 1.8. In Figure 1.9 the approach taken by Ballard Power Systems is shown. The gas is fed in at one corner, and flows over the face of the electrode to the opposite corner, before passing on to the next cell. This diagram is from a 'Proton Exchange Membrane' fuel cell, where the anode, electrolyte and cathode are made in one piece, and are very thin. In this type of cell the bipolar plate makes up the majority of the volume of the fuel cell stack.

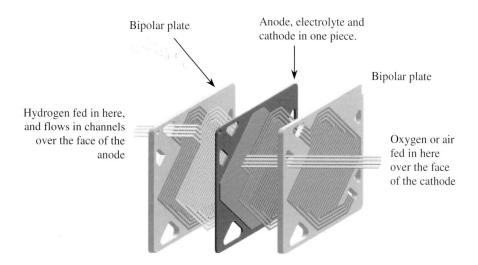

Bipolar plate

Anode, electrolyte and cathode in one piece.

Bipolar plate

Hydrogen fed in here, and flows in channels over the face of the anode

Oxygen or air fed in here over the face of the cathode

Figure 1.9 Diagram showing an "exploded" view of two bipolar plates and the gas flow channels that are formed in their surfaces. This is from a "Proton Exchange Membrane" type fuel cell described briefly below, and in detail in Chapter 4. (Graphic by kind permission of Ballard Power Systems.)

Graphite is a common material for the bipolar plate to be made from. The gas flow grooves have to be carefully machined, for example on a CNC machine, a process which may be quite highly automated, but is still slow and uses an expensive machine. However, there are many other ways of constructing these bipolar plates, some of which are discussed when we look at specific types of fuel cells in later chapters. In some cases the result is not really a "plate" at all, and in these cases the item is called a "cell interconnector". Nevertheless, whatever manufacturing method is used, the cost of the bipolar plates is a very significant part of the cost of a fuel cell system.

Figure 1.10 Photograph of a fuel cell stack under test. The voltage of each of the approximately 60 cells in the stack is being measured. In this type of cell the electrolyte and the electrodes are very thin, and the bulk of the fuel cell is made up of the bipolar plates. (Photograph reproduced by kind permission of Ballard Power Systems)

1.4   Fuel Cell Types

Leaving aside practical issues such as manufacturing and materials cost, the two fundamental technical problems with fuel cells are:-

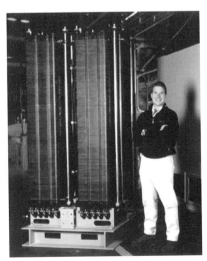

- the slow reaction rate, leading to low currents and power, discussed briefly in Section 1.2 above, and
- that hydrogen is not a readily available fuel.

To solve these problems many different fuel cell types have been tried. The different fuel cell types are usually distinguished by the electrolyte that is used, though there are always other important differences as well. The situation now is that five classes of fuel cell have emerged as viable systems for the present and near future. Basic information about these systems is given in Table 1.1.

Figure 1.11 This photograph illustrates the term "stack" very well. Several hundred cells are connected in series to produce a fuel cell of power about 200 kW. (Photograph reproduced by kind permission of Ballard Power Systems)

Table 1.1. Data for different types of fuel cell.

Fuel Cell Type	Mobile Ion	Operating Temp.	Applications and notes
Alkaline - AFC	OH^-	50 - 200 °C	Used in space vehicles, e.g. Apollo, Shuttle.
Proton exchange membrane (PEM)	H^+	50 - 100 °C	Especially suitable for vehicles and mobile applications, but also for lower power CHP systems
Phosphoric acid PAFC	H^+	~ 220 °C	Large numbers of 200 kW CHP systems in use.
Molten carbonate MCFC	CO_3^{2-}	~650 °C	Suitable for medium to large scale CHP systems, up to MW capacity
Solid oxide SOFC	O^{2-}	500 - 1000 °C	Suitable for all sizes of CHP systems, 2 kW to multi MW.

As well as facing up to different problems, the various fuel types also try to play to the strengths of fuel cells in different ways. The **PEM fuel cell** capitalises on the essential simplicity of the fuel cell. The electrolyte is a solid polymer, in which protons are mobile. The chemistry is the same as the acid electrolyte fuel cell of Figure 1.3. With a solid and immobile electrolyte, this type of cell is inherently very simple.

These cells run at quite low temperatures, so the problem of slow reaction rates is addressed by using sophisticated catalysts and electrodes. Platinum is the catalyst, but developments in recent years mean that only minute amounts are used, and the cost of the platinum is a small part of the total price of a PEM fuel cell. The problem of hydrogen supply is not really addressed – quite pure hydrogen **must** be used, though various ways of supplying this are possible, as is discussed in Chapter 7.

Although PEM fuel cells were used on the first manned spacecraft, the **alkaline fuel cell** was used on the Apollo and Shuttle Orbiter craft. The problem of slow reaction rate is overcome by using highly porous electrodes, with a platinum catalyst, and by operating at quite high pressures. The temperature is higher than for the PEM fuel cell, but it would still normally be classified as a 'low temperature' cell. The alkaline fuel cell is discussed in Chapter 5, where their main problem is described – that the air and fuel supplies must be free from CO_2, or else pure oxygen and hydrogen must be used.

The **phosphoric acid fuel cell (PAFC)** was the first to be produced in commercial quantity and enjoy widespread terrestrial use. Many 200 kW systems, manufactured by the International Fuel Cells Corporation, are installed in the USA and Europe, as well as systems produced by Japanese companies. Porous electrodes, platinum catalysts and a fairly high temperature (~220 °C) are used to boost the reaction rate to a reasonable level. The hydrogen fuel problem is solved by 'reforming' natural gas (CH_4, methane) to hydrogen and carbon dioxide, but the equipment needed to do this adds considerably to the costs, complexity and size of the fuel cell. Nevertheless, they use the inherent simplicity of a fuel cell to provide an extraordinarily reliable and maintenance free power system. They have run continuously for over a year without any maintenance requiring shut-down or human intervention.

Figure 1.12 Phosphoric acid fuel cell. As well as providing 200 kW or electricity it also provides about 200 kW of heat energy in the form of steam. Such units are called "combined heat and power" or CHP systems. (Picture reproduced by kind permission of ONSI Corporation.)

As is the way of things, each fuel cell type solves some problems, but brings new difficulties of its own. The **solid oxide fuel cell (SOFC)** operates in the region of 600 to 1000°C. This means that high reaction rates can be achieved without expensive catalysts, and that gases such as natural gas can be used directly, or 'internally reformed' within the fuel cell, without the need for a separate unit. This fuel cell type thus addresses all the problems and takes full advantage of the inherent simplicity of the fuel cell concept. Nevertheless, the ceramic materials that these cells are made from are difficult to handle, so they are expensive to manufacture, and there is still quite a large amount of extra equipment needed to make a full fuel cell system. This extra plant includes air and fuel pre-heaters, also the cooling system is more complex, and they are not easy to start up.

Despite operating at temperatures of up to 1000°C, the SOFC always stays in the solid state. This is not true for the **molten carbonate fuel cell (MCFC)**, which has the interesting feature that it needs the carbon dioxide in the air to work. The high temperature means that a good reaction rate is achieved using a comparatively inexpensive catalyst – nickel. The nickel also forms the electrical basis of the electrode. Like the SOFC it can use gases such as methane and coal gas (H_2 and CO) directly, without an external reformer. However, this simplicity is somewhat offset by the nature of the electrolyte, a hot and corrosive mixture of lithium, potassium and sodium carbonates.

1.5   Other Parts of a Fuel Cell System

The core of a fuel cell power system is the electrodes, the electrolyte and the bipolar plate that we have already considered. However, other parts frequently make up a large proportion of the engineering of the fuel cell system. These 'extras' are sometimes called the 'balance of plant' (BOP). In the higher temperature fuel cells used in CHP systems the fuel cell stack often appears to be quite a small and insignificant part of the whole system, as is shown in Figure 1.13. The extra components required depend greatly on the type of fuel cell, and crucially on the fuel used. These vitally important subsystems issues are described in much more detail in Chapters 7 to 9, but a summary is given here.

On all but the smallest fuel cells the air and fuel will need to be circulated through the stack using *pumps* or *blowers*. Often *compressors* will be used, which will sometimes be accompanied by the use of *intercoolers*, as in internal combustion engines.[1]

The DC output of a fuel cell stack will rarely be suitable for direct connection to an electrical load, and so some kind of *power conditioning* is nearly always needed. This may be as simple as a voltage regulator, or a **DC/DC converter**. In CHP systems a DC to AC inverter is needed, which is a significant part of the cost of the whole system.[2] **Electric motors** too will nearly always be a vital part of a fuel cell system, driving the pumps, blowers and compressors mentioned above. Frequently also, the electrical power generated will be destined for an electric motor - for example in motor vehicles. The supply and storage of hydrogen is a very critical problem for fuel cells, and it will be discussed in some detail in Chapter 7.

Fuel storage will clearly be a part of many systems. If the fuel cell does not use hydrogen then some form of *fuel processing system* will be needed. These are often very large and complex, for example, when obtaining hydrogen from petrol in a car. In many cases *desulphurisation* of the fuel will be necessary.

Various *control valves* will usually be needed, as well as *pressure regulators*. In most cases a *controller* will be needed to co-ordinate the parts of the system. A special problem the controller has to deal with is the start-up and shut-down of the fuel cell system, as this can be a complex process, especially for the high temperature cells.

For all but the smallest fuel cells a *cooling system* will be needed. In the case of combined heat and power systems, this will usually be called a '*heat exchanger*', as the

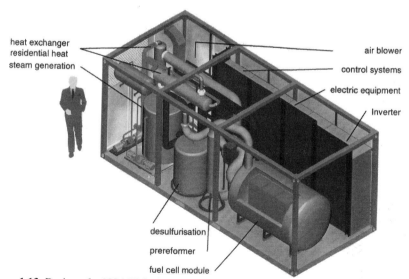

Figure 1.13 Design of a 100 kW fuel cell based combined heat and power system. (Diagram reproduced by kind permission of Siemens Power Generation)

---

[1] These components are discussed in some detail in Chapter 8.
[2] Electrical subsystems are covered in Chapter 9.

idea is not to lose the heat, but to use it somewhere else. Sometimes, in the case of the higher temperature cells, some of the heat generated in the fuel cell will be used in fuel and/or air *pre-heaters*.

This idea of the balance of plant is illustrated in Figures 1.13 and 1.14. In Figure 1.13 we see that the fuel cell module is, in terms of size, a small part of the overall system, which is dominated by the fuel and heat processing systems. This will nearly always be the case for combined heat and power systems running on ordinary fuels such as natural gas. Figure 1.14 is the fuel cell engine from a car. It uses hydrogen fuel, and the waste heat is only used to warm the car interior. The fuel cell stacks are the rectangular blocks to the right. The rest of the unit is much less bulky, but still takes up over half the volume of the whole system.

Figure 1.14   The 60 kW (approx.) fuel cell system for a prototype Mercedes Benz A-class car. (Photograph reproduced by kind permission of Ballard Power Systems)

1.6   Figures used to Compare Systems

When comparing fuel cells with each other, and with other electric power generators, certain standard key figures are used. For comparing fuel cell electrodes and electrolytes the key figure is the current per unit area, always known as the current density. This is usually given in $mA.cm^{-2}$ though some Americans use $A.ft^{-2}$. (Both figures are in fact quite similar: $1.0\ mA.cm^{-2} = 0.8\ A.ft^{-2}$.)

This figure should be given at a specific operating voltage, typically about 0.6 or 0.7 volts. These two numbers can then be multiplied to give the power per unit area, typically given in $mW.cm^{-2}$.

A note of warning should be given here. Electrodes frequently do not 'scale-up' properly. That is, if the area is doubled the current will often *not* double. The reasons for this are varied and often not well understood, but relate to issues such as the even delivery of reactants and removal of products from all over the face of the electrode.

Bipolar plates will be used to connect many cells in series. To the fuel cell stack will be added the 'balance of plant' components mentioned in Section 1.5. This will give a system of a certain power, mass and volume. These figures give the key figures of merit for comparing electrical generators - specific power and power density.

$$Power\ Density\ = \frac{Power}{Volume}$$

The most common unit is $kW.m^{-3}$, though $kW/litre$ is also used.

The measure of power per unit mass is called the specific power:-

$$Specific\quad Power\ = \frac{Power}{Mass}$$

The straightforward SI unit of $W.kg^{-1}$ is used for specific power.

The cost of a fuel cell system is obviously vital, and this is usually quoted in *US dollars per kW*, for ease of comparison.

The lifetime of a fuel cell is rather difficult to specify. Standard engineering measures such as MTBF (mean time between failures) do not really apply well, as a fuel cell's performance always gradually deteriorates, and their power drops fairly steadily with time as the electrodes and electrolyte age. This is sometimes given as the '*percentage deterioration per hour*'. The gradual decline in voltage is also sometimes given in units of mV/1000 hours. Formally, the life is over when it can no longer deliver the rated power, i.e. when a "10 kW fuel cell" can no longer deliver 10 kW. When new the system may have been capable of, say, 25% more than the rated power - 12.5 kW in this case.

The final figure of key importance is the efficiency, though as is explained in the next chapter, this is not at all a straightforward figure to give, and any information needs treating with caution.

In the automotive industry the two key figures are the cost per kW and the power density. In round figures, current internal combustion engine technology is about 1 kW/litre and $10 per kW. Such a system should last about 4000 hours. (That is, about 1 hour use each day for over 10 years). For combined heat and power systems the cost is still important, but a much higher figure of $1000 per kW is the target. The cost is raised by the extra heat exchanger and mains grid connection systems, which are also needed by rival technologies, and because the system must withstand much more constant usage - 40,000 hours use would be a minimum.

1.7   Advantages and Applications

The most important disadvantage of fuel cells at the present time is the same for all types – the cost. However, there are many advantages, which feature more or less strongly for different types and lead to different applications. These include:

- Efficiency. As is explained in the following chapter, fuel cells are generally more efficient than combustion engines whether piston or turbine based. A further feature of this is that small systems can be just as efficient as large ones. This is very important in the case of the small local power generating systems needed for combined heat and power systems.

- Simplicity. The essentials of a fuel cell are very simple, with few if any moving parts. This can lead to highly reliable and long lasting systems.

- Low emissions. The by-product of the main fuel cell reaction, when hydrogen is the fuel, is pure water, which means a fuel cell can be essentially "zero emission". This is their main advantage when used in vehicles, as there is a requirement to reduce vehicle emissions, and even eliminate them within cities. However, it should be noted that, at present, emissions of CO_2 are nearly always involved in the production of the hydrogen needed as the fuel.

- Silence. Fuel cells are very quiet, even those with extensive extra fuel processing equipment. This is very important in both portable power applications and for local power generation in combined heat and power schemes.

The fact that hydrogen is the preferred fuel in fuel cells is, in the main, one of their principal disadvantages. However, there are those who hold that this is a major advantage. It is envisaged that as fossil fuels run out, hydrogen will become the major world fuel and energy vector. It would be generated, for example, by massive arrays of solar cells electrolysing water. This may well be true, but is unlikely to come to pass within the lifetime of this book!

The advantages of fuel cells impact particularly strongly on **combined heat and power** systems (for both large and small scale applications), and on **mobile power systems**, especially for **vehicles** and electronic equipment such as **portable computers, mobile telephones, and military communications equipment.** These areas are the major fields where fuel cells are being used. Several example applications are given in the chapters where the specific fuel cell types are described - especially Chapters 4 and 6. A key point is the very wide range of applications of fuel cell power, from systems of a few watts up to megawatts. In this respect fuel cells are quite unique as energy converters - their range of application far exceeds all others.

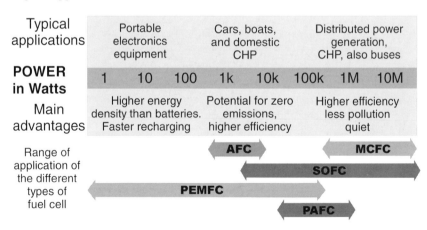

Figure 1.15   Chart to summarise the applications and main advantages of fuel cells of different types, and in different applications.

2

Efficiency and Open Circuit Voltage

In this chapter we consider the **efficiency of fuel cells** - how it is calculated and what the limits are. The energy considerations give us information about the **open circuit voltage** (OCV) of a fuel cell, and the formulas produced also give important information about how factors such as pressure, gas concentration and temperature affect the voltage.

2.1 Energy and the EMF of the Hydrogen Fuel Cell

In some electrical power generating devices it is very clear what form of energy is being converted into electricity. A good example is a wind driven generator, as in Figure 2.1. The energy source is clearly the kinetic energy of the air moving over the blades.

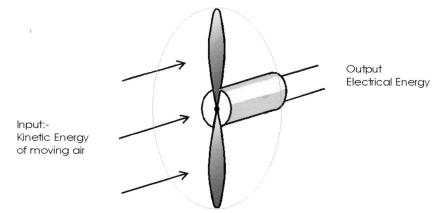

Input:-
Kinetic Energy
of moving air

Output
Electrical Energy

Figure 2.1 A wind driven turbine. The input and output powers are simple to understand and calculate

With a fuel cell such energy considerations are much more difficult to visualise. The basic operation has already been explained, and the input and outputs are shown in Figure 2.2 overleaf.

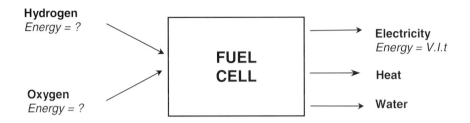

Figure 2.2   Fuel cell inputs and outputs

The electrical power and energy output are easily calculated from the well known formulas:-

$$Power = VI \qquad and \qquad Energy = VIt$$

However, the energy of the chemical inputs and output is not so easily defined. At a simple level we could say that it is the "chemical energy" of the H_2, O_2 and H_2O that is in question. The problem is that "chemical energy" is not simply defined - and terms such as enthalpy, Helmholtz function, and Gibbs free energy are used, as well as older (but still useful) terms such as calorific value.

In the case of fuel cells it is the "Gibbs free energy" that is important. This can be defined as the "energy available to do external work, neglecting any work done by changes in pressure and/or volume". In a fuel cell the "external work" involves moving electrons round an external circuit - any work done by a change in volume between the input and output is not harnessed by the fuel cell[1]. The Helmholtz function is, simply put, the Gibbs free energy less the Pressure × Volume. Enthalpy is the Gibbs free energy plus the energy connected with the entropy (see Appendix 1).

All these forms of "chemical energy" are rather like ordinary mechanical "potential energy" in two important ways.

The **first** is that the *point of zero energy can be defined as almost anywhere*. When working with chemical reactions the zero energy point is normally defined as pure elements, in the normal state, at standard temperature and pressure (25°C, 0.1 MPa). The term "Gibbs free energy of formation", G_f rather than the "Gibbs free energy" is used when adopting this convention. Similarly, we also use "enthalpy of formation" rather than just "enthalpy". For an ordinary hydrogen fuel cell operating at STP[2] this means that the "Gibbs free energy of formation" of the input is zero – a useful simplification.

The **second** parallel with mechanical potential energy is that it is the *change* in energy that is important. In a fuel cell it is the change in this Gibbs free energy of formation,

---

[1] Though it may be harnessed by some kind of turbine in a combined cycle system, as discussed in Chapter 6.

[2] Standard temperature and pressure, or Standard reference state i.e. 100 kPa and 25°C or 298.15K

ΔG_f that gives us the energy released. This change is the difference between the Gibbs free energy of the products and the Gibbs free energy of the inputs or reactants.

$$\Delta G_f = G_f \text{ of products } - G_f \text{ of reactants}$$

To make comparisons easier, it is nearly always most convenient to consider these quantities in their "per mole" form. These are indicated by a ‾ over the lower case letter, e.g. $\left(\overline{g}_f\right)_{H_2O}$ is the molar specific Gibbs free energy of formation for water.

Consider the basic reaction for the hydrogen/oxygen fuel cell:

$$2H_2 + O_2 \rightarrow 2H_2O$$

which is equivalent to:

$$H_2 + \tfrac{1}{2}O_2 \rightarrow H_2O$$

The "product" is one mole of H_2O, and the "inputs" are one mole of H_2 and half a mole of O_2. Thus

$$\Delta g_f = g_f \text{ of products } - g_f \text{ of reactants}$$

So we have

$$\Delta \overline{g}_f = \left(\overline{g}_f\right)_{H_2O} - \left(\overline{g}_f\right)_{H_2} - \tfrac{1}{2}\left(\overline{g}_f\right)_{O_2}$$

Moles - g mole and kg mole

The *mole* is a measure of the "amount" of a substance that takes into account its molar mass. The molar mass of H_2 is 2.0 atomic mass units, so one *gmole* is 2.0g, and one *kgmole* is 2.0kg. Similarly, the molecular weight of H_2O is 18 amu, so 18g is one *gmole*, or 18kg is one *kgmole*. The *gmole*, despite the SI preference for kg, is still the most commonly used, and the "unprefixed" *mole* means *gmole*.

A *mole* of any substance always has the same number of entities (e.g. molecules) - 6.022×10^{23} - called Avagadro's number. This represented by the letter N or N_a.

A "mole of electrons" is 6.022×10^{23} electrons. The charge is $N.e$, where e is 1.602×10^{-19} coulombs - the charge on one electron. This quantity is called the Faraday constant, and designated by the letter F.

$$F = N.e = 96485 \text{ Coulombs}$$

This equation seems straightforward and simple enough. However, the Gibbs free energy of formation is *not constant*, but changes with temperature and state (liquid or gas). Table 2.1 below shows $\Delta \bar{g}_f$ for the basic hydrogen fuel cell reaction

$$H_2 + \tfrac{1}{2}O_2 \rightarrow H_2O$$

for a number of different conditions. The method for calculating these values is given in Appendix 1. Note that the values are negative, which means that energy is released.

Table 2.1 $\Delta \bar{g}_f$ for the reaction

$H_2 + \tfrac{1}{2}O_2 \rightarrow H_2O$ at various temperatures

Form of water product	Temperature (°C)	Δg_f (kJ/mole)
Liquid	25	-237.2
Liquid	80	-228.2
Gas	80	-226.1
Gas	100	-225.2
Gas	200	-220.4
Gas	400	-210.3
Gas	600	-199.6
Gas	800	-188.6
Gas	1000	-177.4

If there are no losses in the fuel cell, or as we should more properly say, if the process is "reversible", then **all** this Gibbs free energy is converted into electrical energy. (In practice, some is also released as heat.) We will use this to find the reversible open circuit voltage of a fuel cell.

The basic operation of a fuel cell was explained in Chapter 1. A review of this Chapter will remind you that, for the hydrogen fuel cell, **two** electrons pass round the external circuit for each water molecule produced and each molecule of hydrogen used. So, for one *mole* of hydrogen used $2N$ electrons pass round the external circuit - where N is Avagadro's number. If $-e$ is the charge on one electron, then the charge that flows is

$$-2Ne = -2F \quad \text{Coulombs}$$

F being the Faraday constant, or the charge on one mole of electrons.

If E is the voltage of the fuel cell, then the electrical work done moving this charge round the circuit is:-

$$\text{Electrical work done} = \text{charge} \times \text{voltage} = -2FE \quad \text{Joules}$$

If the system is reversible (or has no losses) then this electrical work done will be equal to the Gibbs free energy released $\Delta \overline{g}_f$. So

$$\Delta \overline{g}_f \;=\; -2F.E$$

Thus:-

$$E \;=\; \frac{-\Delta \overline{g}_f}{2F} \qquad\qquad [2.1]$$

This fundamental equation gives the EMF or reversible open circuit voltage of the hydrogen fuel cell.

Reversible processes, irreversibilities and losses - an explanation of terms

An example of a simple reversible process is that shown in Fig.2.3. In position ① the ball has no kinetic energy, but potential energy $m.g.h$. In position ② this PE has been converted into kinetic energy. If there is no rolling resistance or wind resistance the process is *reversible*, in that the ball can roll up the other side and recover its potential energy.

In practice some of the potential energy will be converted into heat, because of friction and wind resistance. This is an *irreversible* process as the heat cannot be converted back into kinetic or potential energy. We might be tempted to describe this as a "loss" of energy, but that would not be very precise. In a sense, the potential energy is no more "lost" to heat than it is "lost" to kinetic energy. The difference is that in one you can get it back - it's reversible; and in the other you can't – it's irreversible. So, the term "irreversible energy loss" or "irreversibility" is a rather more precise description of situations that many would describe as simply "a loss".

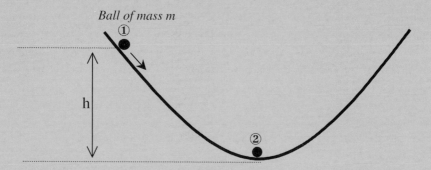

Figure 2.3 Simple reversible process

For example, a hydrogen fuel cell operating at 200°C has $\Delta \overline{g}_f = -220$ kJ, so

$$E = \frac{220000}{2 \times 96485} = 1.14 \ volts$$

Note that this figure assumes no "irreversibilities", and as we shall see later assumes pure hydrogen and oxygen at standard pressure (0.1MPa). In practice the voltage would be lower than this because of the voltage drops discussed in Chapter 3. Some of these irreversibilities apply a little even when no current is drawn, so even the open circuit voltage of a fuel cell will usually be lower than the figure given by equation 2.1.

2.2   The Open Circuit Voltage of other Fuel Cells and Batteries

The equation that we have derived for the open circuit voltage of the hydrogen fuel cell:-

$$E = \frac{-\Delta \overline{g}_f}{2F}$$

can be applied to other reactions too. The only step in the derivation that was specific to the hydrogen fuel cell was the "2" electrons for each molecule of fuel, which led to the 2 in the equation. If we generalise it to any number of electrons per molecule, we have the formula:-

$$E = \frac{-\Delta \overline{g}_f}{zF} \qquad [2.2]$$

where z is the number of electrons transferred for each molecule of fuel.
The derivation was also not specific to fuel cells, and applies just as well to other electro-chemical power sources, particularly primary and secondary batteries. For example, the reaction in the familiar alkali battery used in radios and other portable appliances can be expressed (e.g. Bockris, 1981) by the equation:-

$$2\,MnO_2 \ + \ Zn \quad \rightarrow \quad ZnO \ + \ Mn_2O_3$$

For which $\Delta \overline{g}_f$ is -277 kJ.mol^{-1}. At the anode the reaction is:-

$$Zn \ + \ 2OH^- \quad \rightarrow \quad ZnO \ + \ H_2O \ + \ 2e^-$$

And at the cathode we have:-   Electrons flow from anode to
                                                          cathode

$$2MnO_2 \ + \ H_2O \ + \ 2e^- \quad \rightarrow \quad Mn_2O_3 \ + \ 2OH^-$$

Thus two electrons are passed round the circuit, and so the equation for the open circuit voltage is exactly the same as 2.1. This gives:-

$$E = \frac{2.77 \times 10^5}{2 \times 96485} = 1.44 \; Volts$$

Another useful example is the methanol fuel cell, which we look at later in chapters. The overall reaction is:-

$$2\,CH_3OH \;+\; 3\,O_2 \;\rightarrow\; 4\,H_2O \;+\; 2\,CO_2$$

with 12 electrons passing from anode to cathode, i.e. 6 electrons for each molecule of methanol. For methanol reaction $\Delta \overline{g}_f$ is -698.2 kJ.mol^{-1}. Substituting these numbers into equation 2.2 gives:-

$$E \;=\; \frac{6.98 \times 10^5}{6 \times 96485} = 1.21 \; Volts$$

We note that this is very similar to the open circuit reversible voltage for the hydrogen fuel cell.

2.3   Efficiency and Efficiency Limits

It is not straightforward to define the efficiency of a fuel cell, and efficiency claims cannot usually be taken at face value. In addition, the question of the maximum possible efficiency of a fuel cell is not without its complications.

The wind driven generator of Figure 2.1 is an example of a system where the efficiency is fairly simple to define, and it is also clear that there must be some limit to the efficiency. Clearly, the air passing through the circle defined by the turbine blades cannot lose all its kinetic energy. If it did it would stop dead and there would be an accumulation of air behind the turbine blades. As is explained in books on wind power, it can be shown, using fluid flow theory, that

Maximum energy from generator = $0.58 \times$ kinetic energy of the wind

The figure of 0.58 is known as the "Betz Coefficient".

A more well known example of an efficiency limit is that for heat engines - such as steam and gas turbines. If the maximum temperature of the heat engine is T_1, and the heated fluid is released at temperature T_2, then Carnot showed that the maximum efficiency possible is:

$$\frac{T_1 - T_2}{T_1}$$

The temperatures are in Kelvin, where "room temperature" is about 290K, and so T_2 is never likely to be small. As an example, for a steam turbine operating at 400°C (675K), with the water exhausted through a condenser at 50° (325K), the Carnot efficiency limit is:

$$\frac{675 - 325}{675} = 0.52 = 52\%$$

The reason for this efficiency limit for heat engines is not particularly mysterious. Essentially there must be some heat energy, proportional to the lower temperature T_2, which is always "thrown away" or wasted. This is similar to the kinetic energy that must be retained by the air that passes the blades of the wind driven generator.

With the fuel cell the situation is not so clear. It is quite well known that fuel cells are not subject to the Carnot efficiency limit. Indeed it is commonly supposed that if there were no "irreversibilities" then the efficiency could be 100%, and if we define efficiency in a particular (not very helpful) way, then this is true.

In Section 2.1 we saw that it was the "Gibbs free energy" that is converted into electrical energy. If it were not for the irreversibilities to be discussed in Chapter 3, all this energy would be converted into electrical energy, and the efficiency could be said to be 100%. However, this is the result of choosing one among several types of "chemical energy". We have also noted, in Table 2.1, that the Gibbs free energy changes with temperature, and we will see in the section following that it also changes with pressure and other factors. All in all, to define efficiency as:

$$\frac{\text{electrical energy produced}}{\text{Gibbs free energy change}}$$

is not very useful, and is rarely done, as whatever conditions are used the efficiency limit is always 100%.

Since a fuel cell uses materials that are usually burnt to release their energy, it would make sense to compare the electrical energy produced with the heat that would be produced by burning the fuel. This is sometimes called the "calorific value", though a more precise description is the change in "enthalpy of formation". Its symbol is $\Delta \bar{h}_f$. As with the Gibbs free energy, the convention is that $\Delta \bar{h}_f$ is negative when energy is released. So to get a good comparison with other fuel using technologies, the efficiency of the fuel cell is usually defined as:

$$\frac{\text{electrical energy produced per mole of fuel}}{-\Delta \bar{h}_f} \qquad [2.3]$$

However, even this is not without its ambiguities, as there are two different values that we can use for $\Delta \bar{h}_f$. For the "burning" of hydrogen:

$$H_2 + \tfrac{1}{2}O_2 \rightarrow H_2O \quad (steam)$$
$$\Delta \bar{h}_f = -241.83 \quad kJ / mole$$

Whereas if the product water is condensed back to liquid, the reaction is:

$$H_2 + \tfrac{1}{2}O_2 \rightarrow H_2O \quad (liquid)$$
$$\Delta \bar{h}_f = -285.84 \quad kJ / mole$$

The difference between these two values for $\Delta \bar{h}_f$ (44.01 kJ/mole) is the molar enthalpy of vaporisation[3] of water. The higher figure is called the "higher heating value" (HHV), and the lower, quite logically, the "lower heating value" (LHV). Any statement of efficiency should say whether it relates to the higher or lower heating value. If this information is not given, the LHV has probably been used, since this will give a higher efficiency figure.

We can now see that there is a limit to the efficiency, if we define it as in equation 2.3. The maximum electrical energy available is equal to the change in Gibbs free energy, so:

$$\text{Maximum efficiency possible} = \frac{\Delta \bar{g}_f}{\Delta \bar{h}_f} \times 100\% \qquad [2.4]$$

This maximum efficiency limit is sometimes known as the 'thermodynamic efficiency'. Table 2.2 gives the values of the efficiency limit, relative to the higher heating value, for a hydrogen fuel cell. The maximum voltage, from equation 2.1, is also given.

Table 2.2 $\Delta \bar{g}_f$, maximum EMF, and efficiency limit (ref. HHV) for hydrogen fuel cells

Form of water product	Temp °C	$\Delta \bar{g}_f$, kJ/mole	Max EMF	Efficiency limit
Liquid	25	-237.2	1.23V	83%
Liquid	80	-228.2	1.18V	80%
Gas	100	-225.3	1.17V	79%
Gas	200	-220.4	1.14V	77%
Gas	400	-210.3	1.09V	74%
Gas	600	-199.6	1.04V	70%
Gas	800	-188.6	0.98V	66%
Gas	1000	-177.4	0.92V	62%

---

[3] This used to be known as the molar 'latent heat'.

Figure 2.4 (below) shows how these values vary with temperature, and how they compare with the "Carnot limit". Three important points should be noted:

- Although the graph and table would suggest that lower temperatures are better, the voltage losses discussed in Chapter 3 are nearly always less at higher temperatures. So, *in practice,* fuel cell voltages are usually *higher* at higher temperatures.

- The waste heat from the higher temperature cells is more useful than that from lower temperature cells.

- Contrary to statements often made by their supporters, fuel cells do NOT always have a higher efficiency limit than heat engines.[4]

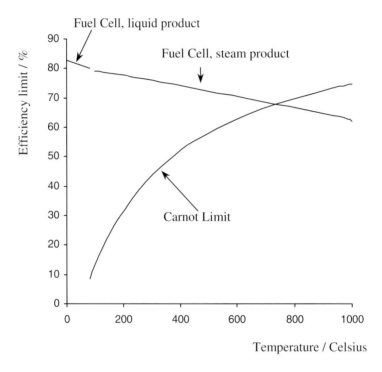

Figure 2.4 Maximum H_2 fuel cell efficiency at standard pressure, referred to Higher Heating Value. The Carnot limit is shown for comparison, with 50°C exhaust temperature

This decline in maximum possible efficiency with temperature associated with the hydrogen fuel cell does not occur in exactly the same way with other types of fuel cell. For example, when using carbon monoxide:

$$CO + \tfrac{1}{2}O_2 \rightarrow CO_2$$

---

[4] In Chapter 6 we see how a heat engine and a high temperature fuel cell can be combined into a particularly efficient system.

$\Delta \overline{g}_f$ changes even more quickly with temperature, and the maximum possible efficiency falls from about 82% at 100°C to 52% at 1000°C. On the other hand for the reaction:

$$CH_4 + 2O_2 \rightarrow CO_2 + 2H_2O$$

$\Delta \overline{g}_f$ is fairly constant with temperature, and the maximum possible efficiency hardly changes.

2.4   Efficiency and the Fuel Cell Voltage

It is clear from Table 2.2 that there is a connection between the maximum EMF of a cell and its maximum efficiency. The operating voltage of a fuel cell can also be very easily related to its efficiency. This can be shown by adapting equation 2.1. If all the energy from the hydrogen fuel, its "calorific value", heating value, or enthalpy of formation, were transformed into electrical energy, then the EMF would be given by:

$$E = \frac{-\Delta \overline{h}_f}{2F}$$

$$= 1.48 \text{ volts if using the HHV}$$
$$\text{or} \quad = 1.25 \text{ volts if using the LHV}$$

These are the voltages that would be obtained from a 100% efficient system, referred to the HHV or LHV. The actual efficiency of the cell is then the actual voltage divided by these values, or

$$Cell\ efficiency = \frac{V_c}{1.48} 100\% \ \left(ref.\ HHV\right)$$

However, there is a proviso we need to add to this, and that is the issue of fuel usage. In practice, not all the fuel that is input into a fuel cell is used, and some always has to be passed through. A *fuel utilisation coefficient* can be defined as:

$$\mu_f = \frac{mass\ of\ fuel\ reacted\ in\ cell}{mass\ of\ fuel\ input\ to\ cell}$$

The fuel cell efficiency is therefore given by:

$$Efficiency,\ \eta = \mu_f \frac{V_c}{1.48} 100\% \qquad\qquad [2.5]$$

If a figure relative to the LHV is required, use 1.25 instead of 1.48. A good estimate for μ_f is 0.95, which allows the efficiency of a fuel cell to be accurately estimated from the very simple measurement of its voltage. However, it can be a great deal less in some circumstances, as is discussed in Section 2.5.3 below, and Chapter 6.

2.5   The Effect of Pressure and Gas Concentration

2.5.1   The Nernst equation

In Section 2.1 we noted that the Gibbs free energy changes in a chemical reaction vary with temperature. Equally important, though more complex, are the changes in Gibbs free energy with reactant pressure and concentration.
    Consider a general reaction such as:-

$$jJ \quad + \quad kK \rightarrow \quad mM \qquad\qquad\qquad [2.6]$$

where j moles of J react with k moles of K to produce m moles of M. Each of the reactants, and the products, have an associated "activity". This "activity" is designated by a, a_J and a_K being the activity of the reactants, and a_M the activity of the product. It is beyond the scope of this book to give a thorough description of "activity". However, for the case of gases behaving as "ideal gases" then it can be shown that

$$activity \quad a = \frac{P}{P^0}$$

Where P is the pressure or partial pressure of the gas and P^0 is standard pressure, 0.1 MPa. Since fuel cells are generally gas reactors, this simple equation is thus very useful. We can say that activity is proportional to partial pressure. In the case of dissolved chemicals the activity is linked to the molarity (strength) of the solution. The case of the water produced in fuel cells is somewhat difficult, since this can either be as steam or liquid. For steam, we can say that:-

$$a_{H_2O} = \frac{P_{H_2O}}{P^0_{H_2O}}$$

where $P^0_{H_2O}$ is the vapour pressure of the steam at the temperature concerned. This has to be found from steam tables. In the case of liquid water product it is a reasonable approximation to assume that $a_{H_2O} = 1$.

    The activities of the reactants and products modify the Gibbs free energy change of a reaction. Using thermodynamic arguments (e.g. Balmer, 1990), it can be shown that, in a chemical reaction such as that given in 2.6

$$\Delta \bar{g}_f = \Delta \bar{g}_f^0 - RT \ln \left(\frac{a_J^j . a_K^k}{a_M^m} \right)$$

where $\Delta \bar{g}_f^0$ is the change in molar Gibbs free energy of formation at standard pressure.

Despite not looking very "friendly", this equation is useful, and not very difficult. In the case of the hydrogen fuel cell reaction:

$$H_2 + \tfrac{1}{2}O_2 \rightarrow H_2O$$

the equation becomes:

$$\Delta \bar{g}_f = \Delta \bar{g}_f^0 - RT \ln \left(\frac{a_{H_2} . a_{O_2}^{\frac{1}{2}}}{a_{H_2O}} \right)$$

$\Delta \bar{g}_f^0$ is the quantity given in Tables 2.1 and 2.2. We can see that if the activity of the reactants increases, $\Delta \bar{g}_f$ becomes more negative, i.e. more energy is released. On the other hand, if the activity of the product increases, $\Delta \bar{g}_f$ increases, so becomes less negative, and less energy is released. To see how this equation affects voltage we can substitute it into equation 2.1 and obtain:

$$E = \frac{-\Delta \bar{g}_f^0}{2F} + \frac{RT}{2F} \ln \left(\frac{a_{H_2} . a_{O_2}^{\frac{1}{2}}}{a_{H_2O}} \right)$$

$$= E^0 + \frac{RT}{2F} \ln \left(\frac{a_{H_2} . a_{O_2}^{\frac{1}{2}}}{a_{H_2O}} \right) \qquad [2.7]$$

where E^0 is the EMF at standard pressure, and is the number given in column 4 of Table 2.2. The equation shows precisely how raising the activity of the reactants increases the voltage.

Equation 2.7, and its variants below, which give an EMF in terms of product and/or reactant activity, are called "Nernst" equations. The EMF calculated from such equations is known as the "Nernst voltage", and is the reversible cell voltage that would exist at a given temperature and pressure. The logarithmic function involving the reactants allows us to use the regular rules of the logarithmic functions such as:

$$\ln \left(\frac{a}{b} \right) = \ln(a) - \ln(b) \qquad and \qquad \ln \left(\frac{c^2}{d^{\frac{1}{2}}} \right) = 2\ln(c) - \tfrac{1}{2}\ln(d)$$

This makes it straightforward to manipulate equation 2.7 to get at the effect of different parameters. For example, in the reaction:

$$H_2 + \tfrac{1}{2}O_2 \rightarrow H_2O \ (steam)$$

at high temperature (e.g. in a SOFC at 1000°C) we can assume that the steam behaves as an ideal gas, and so

$$a_{H_2} = \frac{P_{H_2}}{P^0}, \qquad a_{O_2} = \frac{P_{O_2}}{P^0}, \qquad a_{H_2O} = \frac{P_{H_2O}}{P^0}$$

Then equation 2.7 will become:

$$E = E^o + \frac{RT}{2F} \ln \left(\frac{\dfrac{P_{H_2}}{P^0} \cdot \left(\dfrac{P_{O_2}}{P^0} \right)^{\frac{1}{2}}}{\dfrac{P_{H_2O}}{P^0}} \right)$$

If all the pressures are given in bar then $P^0 = 1$ and the equation simplifies to:

$$E = E^0 + \frac{RT}{2F} \ln \left(\frac{P_{H_2} \cdot P_{O_2}^{\frac{1}{2}}}{P_{H_2O}} \right) \tag{2.8}$$

In nearly all cases the pressures in equation 2.8 will be partial pressures. That is, the gases will be part of a mixture. For example the hydrogen gas might be part of a mixture of H_2 and CO_2 from a fuel reformer, together with product steam. The oxygen will nearly always be part of air. It is also often the case that the pressure on both the cathode and the anode is approximately the same - this simplifies the design. If this system pressure is P, then we can say that:

$$P_{H_2} = \alpha P$$
$$P_{O_2} = \beta P$$
$$and \quad P_{H_2O} = \delta P$$

where α, β and δ are constants depending on the molar masses and concentrations of H_2, O_2 and H_2O. Equation 2.8 then becomes:

$$E = E^0 + \frac{RT}{2F} \ln\left(\frac{\alpha.\beta^{\frac{1}{2}}}{\delta}.P^{\frac{1}{2}}\right)$$

$$= E^0 + \frac{RT}{2F} \ln\left(\frac{\alpha.\beta^{\frac{1}{2}}}{\delta}\right) + \frac{RT}{4F} \ln(P) \qquad [2.9]$$

The two equations 2.8 and 2.9 are forms of the Nernst equation. They provide a theoretical basis and a quantitative indication for a large number of variables in fuel cell design and operation. Some of these are discussed in later chapters, but some points are considered briefly here.

Partial Pressures

In a mixture of gases the total pressure is made up from the sum of all the "partial pressures" of the components of the mixture. For example, in air at 0.1MPa, the partial pressures are:

Table 2.3 Partial Pressures of atmosphere gases

Gas	Partial Pressure
Nitrogen	0.07809 MPa
Oxygen	0.02095 MPa
Argon	0.00093 MPa
Others (inc. CO_2)	0.00003 MPa
Total	0.10000 MPa

In fuel cells, the partial pressure of oxygen is important. On the fuel side the partial pressure of hydrogen and/or carbon dioxide is important if a hydrocarbon (e.g. CH_4) is used as the fuel source. In such cases the partial pressure will vary according to the reformation method used - see Chapter 7.

However, if the mixture of the gases is known the partial pressures can be easily found. It can be readily shown, using the gas law equation $PV = NRT$, that the volume fraction, molar fraction and pressure fraction of a gas mixture are all equal. For example, consider the reaction:

$$CH_4 + H_2O \rightarrow 3H_2 + CO_2$$

The product gas stream contains 3 parts H_2 and 1 part CO_2 by moles and volume. So, if the reaction takes place at 0.10 MPa:-

$$P_{H_2} = \frac{3}{4} \times 0.1 = 0.075 \text{ MPa} \quad \text{and} \quad P_{CO_2} = \frac{1}{4} \times 0.1 = 0.025 \text{ MPa}$$

2.5.2   Hydrogen partial pressure

Hydrogen can either be supplied pure, or as part of a mixture. If we isolate the pressure of hydrogen term in equation 2.8 we have:

$$ E \;=\; E^0 \;+\; \frac{RT}{2F}\ln\left(\frac{P_{O_2}^{\frac{1}{2}}}{P_{H_2O}}\cdot\right) \;+\; \frac{RT}{2F}\ln\left(P_{H_2}\right) $$

So, if the hydrogen partial pressure changes, say, from P_1 to P_2 bar, with P_{O_2} and P_{H_2O} unchanged, then the voltage will change by:

$$ \Delta V \;=\; \frac{RT}{2F}\ln\left(P_2\right) \;-\; \frac{RT}{2F}\ln\left(P_1\right) $$

$$ =\; \frac{RT}{2F}\ln\left(\frac{P_2}{P_1}\right) \qquad\qquad [2.10] $$

The use of H_2 mixed with CO_2 occurs particularly in Phosphoric acid fuel cells, operating at about 200°C. Substituting the values for R, T and F gives:-

$$ \Delta V \;=\; 0.02\ln\left(\frac{P_2}{P_1}\right) \; volts $$

This gives good agreement with experimental results, which correlate best with a factor of 0.024 instead of 0.020 (Hirschenhofer, 1995). As an example, changing from pure hydrogen to 50% hydrogen/carbon dioxide mixture will reduce the voltage by 0.015 volts per cell.

2.5.3   Fuel and oxidant utilisation

As air passes through a fuel cell the oxygen is used, and so the partial pressure will reduce. Similarly the fuel partial pressure will often decline, as the proportion of fuel reduces and reaction products increase. Referring to equation 2.9 we can see that α and β decrease, whereas δ increases. All these changes make the term

$$ \frac{RT}{2F}\ln\left(\frac{\alpha.\beta^{\frac{1}{2}}}{\delta}\right) $$

from equation 2.9 smaller, and so the EMF will fall. This will vary within the cell - it will be worst near the fuel outlet as the fuel is used. Because of the low resistance bipolar plates on the electrode, it is not actually possible for different parts of one cell to have different voltages, so the current varies. The current density will be lower nearer the exit where the fuel concentration is lower. The RT term in the equation also shows us that this drop in Nernst voltage due to fuel utilisation will be greater in high temperature fuel cells.

We have seen in Section 2.4 above that, for a high system efficiency, the fuel utilisation should be as high as possible. However, this equation shows us that cell voltage, and hence the cell efficiency will *fall* with higher utilisation. So we see fuel and oxygen utilisation need careful optimising, especially in higher temperature cells. The selection of utilisation is an important aspect of system design and is especially important when reformed fuels are used. It is given further consideration in Chapter 6.

2.5.4   System pressure

The Nernst equation in the form of equation 2.9 shows us that the EMF of a fuel cell is increased by the system pressure according to the term:

$$\frac{RT}{4F}\ln(P)$$

So, if the pressure changes from P_1 to P_2 there will be a change of voltage

$$\Delta V = \frac{RT}{4F}\ln\left(\frac{P_2}{P_1}\right)$$

For a SOFC operating at 1000°C, this would give:

$$\Delta V = 0.027\ln\left(\frac{P_2}{P_1}\right)$$

This gives very good agreement with reported results (Bevc, 1997 & Hirschenhofer, 1995) for high temperature cells. However, for other fuel cells, working at lower temperatures, the agreement is not so good. For example, a phosphoric acid fuel cell working at 200°C should be affected by system pressure by the equation:

$$\Delta V = \frac{RT}{4F}\ln\left(\frac{P_2}{P_1}\right) = 0.010\ln\left(\frac{P_2}{P_1}\right)$$

whereas reported results (Hirschenhofer, 1995) give a correlation to the equation

$$\Delta V = 0.063\ln\left(\frac{P_2}{P_1}\right)$$

In other words, at lower temperatures, the benefits of raising system pressure are much greater than the Nernst equation predicts. This is because, except for very high temperature cells, increasing the pressure also reduces the losses at the electrodes, especially the cathode. (This is considered further in Chapter 3.)

A similar effect occurs when studying the change from air to oxygen. This effectively changes β in equation 2.9 from 0.21 to 1.0. Isolating β in equation 2.9 gives:

$$E = E^0 + \frac{RT}{4F}\ln(\beta) + \frac{RT}{2F}\ln\left(\frac{\alpha}{\delta}\right) + \frac{RT}{4F}\ln(P)$$

For the change in β from 0.21 to 1.0, with all other factors remaining constant, we have

$$\Delta V = \frac{RT}{4F}\ln\left(\frac{1.0}{0.21}\right)$$

For a PEM fuel cell at 80°C this would give

$$\Delta V = 0.012 volts$$

In fact, reported results (e.g.Prater, 1990) give a much larger change, 0.05 volts being a typical result. This is also due to the improved performance of the cathode when using oxygen, reducing the voltage losses there.

2.5.5   An application - blood alcohol measurement

As well as generating electrical power, fuel cells are also the basis of some types of sensor. One of the most successful is the fuel cell based alcohol sensor - the "breathalyser". This measures the concentration of alcohol in the air that someone breathes out of their lungs. It has been shown that this is directly proportional to the concentration of alcohol in the

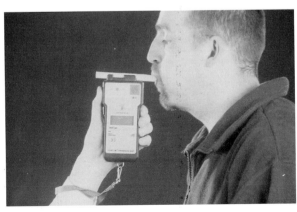

Figure 2.5 Fuel cell based breathalyser as used by police forces in the USA. (photograph supplied by and reproduced by permission of Lion Laboratories Ltd.)

blood. The basic chemistry is that the alcohol (ethanol) is reacted in a simple fuel cell to give a (very small) voltage. In theory the ethanol could be fully oxidised to CO_2 and water. However, the ethanol is probably not fully reacted, and is only partially oxidised to ethanal.

The anode and cathode reactions are probably:-

$$C_2H_5OH \text{ (ethanol)} \rightarrow CH_3CHO \text{ (ethanal)} + 2H^+ + 2e^-$$

And:-
Electrons flow round
external circuit

$$\tfrac{1}{2}O_2 + 2H^+ + 2e^- \rightarrow H_2O$$

We have seen that the voltage is affected by the concentration of the reactants, and so the voltage of the cell will be affected by the concentration of the alcohol in the gas blown into the cell. Thus we can get a measure of the alcohol concentration in the blood. This type of fuel cell is the basis of the majority of roadside blood alcohol measurement instruments used by police forces throughout the world.

2.6   Summary

The reversible open circuit voltage for a hydrogen fuel cell is given by the equation:-

$$E = \frac{-\Delta \bar{g}_f}{2F} \qquad\qquad [2.1]$$

In general, for a reaction where z electrons are transferred for each molecule of fuel the reversible open circuit voltage is:-

$$E = \frac{-\Delta \bar{g}_f}{zF} \qquad\qquad [2.2]$$

However, $\Delta \bar{g}_f$ changes with temperature and other factors. The maximum efficiency is given by the expression:-

$$\eta_{max} = \frac{\Delta \bar{g}_f}{\Delta \bar{h}_f} \times 100\% \qquad\qquad [2.4]$$

The efficiency of a working hydrogen fuel cell can be found from the simple formula:-

$$\text{Efficiency, } \eta = \mu_f \frac{V}{1.48} 100\% \qquad\qquad [2.5]$$

where μ_f is the fuel utilisation (typically about 0.95) and V is the voltage of a single cell within the fuel cell stack. This gives the efficiency relative to the higher heating value of hydrogen.

The pressure and concentration of the reactants affects the Gibbs free energy, and thus the voltage. This is expressed in the Nernst equation, which can be given in many forms. For example if the pressures of the reactants and products are in bar, and the water product is in the form of steam, then:-

$$E = E^0 + \frac{RT}{2F} \ln\left(\frac{P_{H_2} \cdot P_{O_2}^{\frac{1}{2}}}{P_{H_2O}} \right) \qquad [2.8]$$

where E^0 is the cell EMF at standard pressure.

Now, in most of this chapter we have referred to or given equations for the EMF of a cell, or its reversible open circuit voltage. In practice the operating voltage is less than these equations give, and in some cases much less. This is the result of losses or irreversibilities, to which we give careful consideration in the following chapter.

References

Balmer R., (1990) *Thermodynamics*, West,

Bevc F., (1997) "Advances in solid oxide fuel cells and integrated power plants", Proc Instn Mech Engrs, , Vol 211, Part A, pp359-366

Bockris J. O'M, Conway B.E., Yeager E., White R.E. (Eds.), (1981) *A Comprehensive Treatment of Electrochemistry*, Vol. 3, Plenum Press, p220

Hirschenhofer J.H., Stauffer D.B., Engleman R.R., (1995) *Fuel Cells: A Handbook*, Business/Technology Books, p.3-17 and p.5-23

Prater K., (1990) "The renaissance of the Solid Polymer Fuel Cell", *Journal of Power Sources,* Vol. 29, pp 239-250

3

Operational Fuel Cell Voltages

3.1   Introduction

We have seen in Chapter 2 that the theoretical value of the open circuit voltage of a hydrogen fuel cell is given by the formula:-

$$E = \frac{-\Delta \overline{g}_f}{2 F}$$ [3.1]

This gives a value of about 1.2 volts for a cell operating below 100 °C. However, when a fuel cell is made and put to use it is found that the voltage is less than this, often considerably less. Figure 3.1 overleaf shows the performance of a typical single cell operating at about 40 °C, at normal air pressure. The key points to notice about this graph of the cell voltage against current density[1] are:-

- Even the open circuit voltage is less than the theoretical value.
- There is a rapid initial fall in voltage.
- The voltage then falls less slowly, and more linearly.
- There is sometimes a higher current density at which the voltage falls rapidly away.

If a fuel cell is operated at higher temperatures, the shape of the voltage/current density graph changes. As we have seen in Chapter 2, the reversible "no loss" voltage falls. However, the difference between the actual operating voltage and the "no loss" value usually becomes less. In particular the initial fall in voltage as current is drawn from the cell is markedly less. Figure 3.2 below shows the situation for a typical solid oxide fuel cell operating at about 800 °C. The key points here are:-

- The open circuit voltage is equal to or only a very little less than the theoretical value.
- The initial fall in voltage is very small, and the graph is much more linear.
- There may be a higher current density at which the voltage falls rapidly away, as with lower temperature cells.

---

[1] Note that we usually refer to current **density**, or current per unit area, rather than just current. This is to make comparison between cells of different size easier.

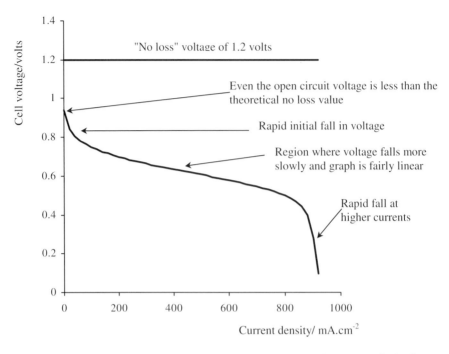

Figure 3.1 Graph showing the voltage for a typical low temperature, air pressure, fuel cell

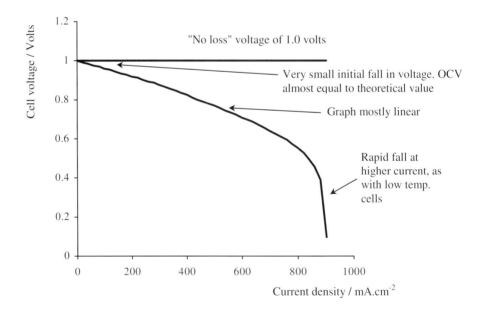

Figure 3.2 Graph showing the voltage of a typical air pressure fuel cell operating at about 800 °C.

Comparing Figures 3.1 and 3.2 we see that although the reversible or "no loss" voltage is lower for the higher temperature, the operating voltage is generally *higher*, because the voltage drop or irreversibilities are smaller.

In this chapter we consider what causes the voltage to fall below the irreversible value, and what can be done to improve the situation.

3.2   Terminology

Fuel cell power systems are highly inter-disciplinary - they require the skills of chemists, electro-chemists, materials scientists, thermodynamicists, electrical and chemical engineers, control and instrumentation engineers, and others, to make them work well. One problem with this is that there are occasions when these various disciplines have their own names for what is often essentially the same thing. The main topic of this chapter is a case in point.

The graphs of Figures 3.1 and 3.2 show the difference between the voltage we would expect from a fuel cell operating ideally (reversibly) and the actual voltage. This difference is the main focus of this chapter. However, a problem is that there are at least *five* commonly used names for this voltage difference!

- **Overvoltage** or **overpotential** is the term often used by electro-chemists. This is because it is a voltage superimposed over the reversible or ideal voltage. Strictly this term only applies to differences generated at an electrode interface. A disadvantage of it is that it tends to imply making the voltage larger - whereas in fuel cells the overvoltage opposes and reduces the reversible ideal voltage.

- **Polarisation** is another term much used by some electro-chemists and others. For many engineers and scientists this has connotations of static electricity, which is not helpful.

- **Irreversibility** is the best term from a thermodynamics point of view. However, it is perhaps too general, and does not connect well with its main effect in this case - reducing a voltage.

- **Losses** is a term that can be used, but is rather too vague. See also the section on "reversibility, irreversibility and losses" in Chapter 2.

- **Voltage drop** is not a very scientifically precise term, but it conveys well the effect observed, and is a term readily understood by electrical engineers.

The English language is rich, frequently having many words for similar ideas. This is an example of how this richness can develop, and most of these terms will be used in this book.

3.3   Fuel Cell Irreversibilities - Causes of Voltage Drop

The characteristic shape of the voltage/current density graphs of Figures 3.1 and 3.2 results from four major irreversibilities. These will be outlined very briefly here, and then considered in more detail in the sections that follow.

1. **Activation losses**. These are caused by the slowness of the reactions taking place on the surface of the electrodes. A proportion of the voltage generated is lost in driving the

chemical reaction that transfers the electrons to or from the electrode. As we shall see in Section 3.4 below, this voltage drop is highly non-linear.

2. **Fuel crossover and internal currents**. This energy loss results from the waste of fuel passing through the electrolyte, and, to a lesser extent, electron conduction through the electrolyte. The electrolyte should only transport ions through the cell, as in Figures 1.3 and 1.4. However, a certain amount of fuel diffusion and electron flow will always be possible. The fuel loss and current is small, and its effect is usually not very important. However, it does have a marked effect on the open circuit voltage of low temperature cells, as we shall see in Section 3.5.

3. **Ohmic losses**. This voltage drop is the straightforward resistance to the flow of electrons through the material of the electrodes and the various interconnections, as well as the resistance to the flow of ions through the electrolyte. This voltage drop is essentially proportional to current density, linear, and so is called "ohmic" losses, or sometimes "resistive" losses.

4. **Mass transport or concentration losses**. These result from the change in concentration of the reactants at the surface of the electrodes as the fuel is used. We have seen in Chapter 2 that concentration affects voltage, and so this type of irreversibility is sometimes called "concentration" loss. Because the reduction in concentration is the result of a failure to transport sufficient reactant to the electrode surface, this type of loss is also often called "mass transport" loss. This type of loss has a third name - "Nernstian". This is because of its connections with concentration, and the effects of concentration are modelled by the Nernst equation.

These four categories of irreversibility are considered one by one in the sections that follow.

3.4   Activation Losses

3.4.1   The Tafel equation

As a result of experiments, rather than theoretical considerations, Tafel observed and reported in 1905 that the overvoltage at the surface of an electrode followed a similar pattern in a great variety of electrochemical reactions. This general pattern is shown in Figure 3.3 below. It shows that if a graph is plotted of overvoltage against *log* of current density, then, for most values of overvoltage, the graph approximates to a straight line. Such plots of overvoltage against *log* of current density are known as "Tafel Plots". The diagram shows two typical plots.

For most values of overvoltage its value is given by the equation:-

$$V = a \log\left(\frac{i}{i_o}\right)$$

This equation is known as the Tafel equation. It can be expressed in many forms. One simple variation is to use natural logarithms instead of base 10, which is preferred since they feature in the equations used in Chapter 2. This gives:-

$$V = A \ \ln\left(\frac{i}{i_o}\right) \qquad [3.2]$$

The constant A is higher for an electrochemical reaction which is slow. The constant i_0 is higher if the reaction is faster. The current density i_0 can be considered as the current density at which the overvoltage begins to move from zero. It is important to remember that the Tafel equation only holds true when $i > i_0$. This current density i_0 is usually called the "exchange" current density, as we shall see in Section 3.4.2.

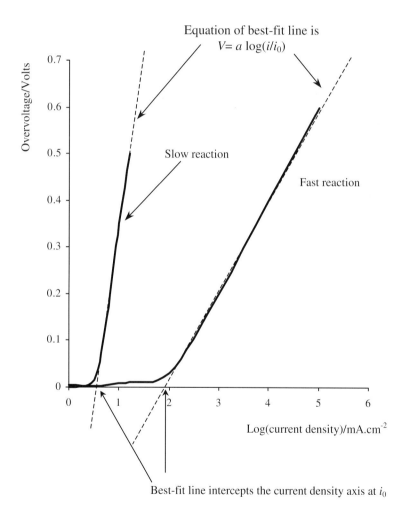

Figure 3.3 Tafel plots for slow and fast electrochemical reactions

3.4.2   The constants in the Tafel equation

Although the Tafel equation was originally deduced from experimental results, it also has a theoretical basis. It can be shown (e.g. McDougall, 1976) that, for a hydrogen fuel cell, the constant A in equation 3.2 above is given by :

$$A = \frac{RT}{2\alpha F}$$ [3.3]

The constant α is called the "charge transfer coefficient", and is the proportion of the electrical energy applied that is harnessed in changing the rate of an electrochemical reaction. Its value depends on the reaction involved and the material the electrode is made from, but it must be in the range 0 to 1.0. For the hydrogen electrode its value is about 0.5 for a great variety of electrode materials (Davies, 1967). At the oxygen electrode the charge transfer coefficient shows more variation, but is still between about 0.1 and 0.5 in most circumstances. In short, experimenting with different materials to get the best possible value for A will make little impact.

The appearance of T in equation 3.3 might give the impression that raising the temperature increases the overvoltage. In fact this is very rarely the case, as the effect of increases in i_0 with temperature far outweigh any increase in A. Indeed we shall see that the key to making the activation overvoltage as low as possible is this i_0, which can vary by several orders of magnitude. Furthermore, it is affected by several parameters other than the material used for the electrode.

The current density i_0 is called the "exchange current density", and it can be visualised as follows. The reaction at the oxygen electrode of a PEM or acid electrolyte fuel cell is:-

$$O_2 \quad + \quad 4e^- \quad + \quad 4H^+ \quad \rightarrow \quad 2H_2O$$

At zero current density we might suppose that there was no activity at the electrode, and that this reaction does not take place. In fact this is not so, the reaction is taking place all the time, but the reverse reaction is also taking place at the same rate. There is an equilibrium expressed as:-

$$O_2 \quad + \quad 4e^- \quad + \quad 4H^+ \quad \leftrightarrow \quad 2H_2O$$

Thus there is a continual backwards and forwards flow of electrons from and to the electrolyte. This current density is i_0, the exchange current density. It is self evident that if this current density is high, then the surface of the electrode is more "active", and a current in one particular direction is more likely to flow. We are simply shifting in one particular direction something already going on, rather than starting something new.

This exchange current density i_0 is crucial in controlling the performance of a fuel cell electrode. It is vital to make its value as high as possible.

Imagine a fuel cell which has no losses at all except for this activation overvoltage on *one* electrode. Its voltage would then be given by the equation:-

$$V = E \;-\; A \ln\left(\frac{i}{i_o}\right) \qquad\qquad [3.4]$$

Where E is the reversible open circuit voltage given by equation 3.1. If we plot graphs of this equation using values of i_0 of 0.01, 1.0 and 100 mA, using a typical value for A of 0.06 volts, we get the curves shown in Figure 3.4 below.

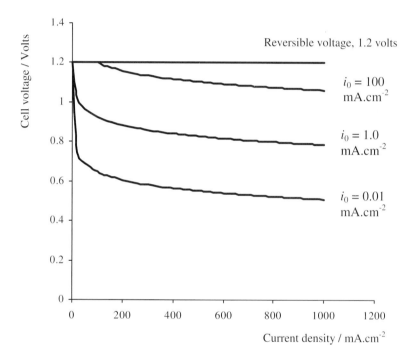

Figure 3.4 Graphs of cell voltage against current density, assuming losses are due only to the activation overvoltage at one electrode, for three different values of exchange current density i_0.

The importance of i_0 can be clearly seen. The effect, for most values of current density, is to reduce the cell voltage by a fairly fixed amount, as we could predict from the Tafel equation. The smaller is i_0, the greater is this voltage drop. Note that when i_0 is 100 mA.cm^{-2}, then there is no voltage drop until the current density i is great than 100mA.cm^{-2}.

It is possible to measure this overvoltage at each electrode, either using reference electrodes within a working fuel cell, or using half-cells. Table 3.1 below gives the values of i_0 for the hydrogen electrode at 25 °C, for various metals. The measurements are for flat smooth electrodes.

Table 3.1 i_0 for the hydrogen electrode
for various metals, for an acid
electrolyte. (Bloom, 1981)

Metal	$i_0 (A.cm^{-2})$
Pb	2.5×10^{-13}
Zn	3×10^{-11}
Ag	4×10^{-7}
Ni	6×10^{-6}
Pt	5×10^{-4}
Pd	4×10^{-3}

The most striking thing about these figures is their great variation, indicating a strong catalytic effect. The figures for the oxygen electrode also vary greatly, and are generally lower by a factor of about 10^5 – much smaller (Appleby & Foulkes, 1993). This would give a figure that is about 10^{-8} A.cm^{-2}, even using Pt catalyst, far worse than even the lowest curve on Figure 3.4. However, the value of i_0 for a real fuel cell electrode is much higher than the figures in Table 3.1, because of the roughness of the electrode. This makes the real surface area many times bigger, typically at least 10^3 times larger, than the nominal length × width.

We have noted that i_0 at the oxygen electrode (the cathode) is much smaller than that at the hydrogen anode, sometimes 10^5 times smaller. Indeed, it is generally reckoned that the overvoltage at the anode is negligible compared to that of the cathode, at least in the case of hydrogen fuel cells. For a low temperature, hydrogen fed fuel cell running on air at ambient pressure, a typical value for i_0 would be about 0.1 mA.cm^{-2} at the cathode, and about 200 mA.cm^{-2} at the anode.

In other fuel cells, for example the direct methanol fuel cell, the anode overvoltage is by no means negligible. In these cases the equation for the total activation overvoltage would combine the overvoltages at both anode and cathode, giving:-

$$Activation\ Voltage\ drop = A_a \ln\left(\frac{i}{i_{0a}}\right) + A_c \ln\left(\frac{i}{i_{0c}}\right)$$

However, it is readily proved that this equation can be expressed as:-

$$V = A \ln\left(\frac{i}{b}\right)$$

where:-

$$A = A_a + A_c \qquad and \qquad b = i_{0a}^{\frac{A_a}{A}} + i_{0c}^{\frac{A_c}{A}} \qquad [3.5]$$

This is exactly the same form as equation 3.2, the overvoltage for one electrode. So whether the activation overvoltage arises mainly at one electrode only, or both, the equation that models the voltage is of the same form. Furthermore, in all cases, the item in the equation that shows the most variation is the exchange current density i_0, rather than A.

3.4.3   Reducing the activation overvoltage

We have seen that the exchange current density i_0 is the crucial factor in reducing the activation overvoltage. A crucial factor in improving fuel cell performance is therefore to increase the value of i_0, especially at the cathode. This can be done in the following ways:-

- Raising the cell temperature. This fully explains the different shape of the voltage/current density graphs of low and high temperature fuel cells illustrated in Figures 3.1 and 3.2. For a low temperature cell i_0 at the cathode will be about 0.1 mA, whereas for a typical 800 °C cell it will be about 10 mA, a 100-fold improvement.
- Using more effective catalysts. The effect of different metals in the electrode is shown clearly by the figures in Table 3.1.
- Increasing the roughness of the electrodes. This increases the real surface area of each nominal 1 cm², and this increases i_0.
- Increasing reactant concentration, e.g. using pure O_2 instead of air. This works because the catalyst sites are more effectively occupied by reactants. (As we have seen in Chapter 2, this also increases the reversible open circuit voltage.)
- Increasing the pressure. This is also presumed to work by increasing catalyst site occupancy. (This also increases the reversible open circuit voltage, and so brings a "double benefit".)

Increasing the value of i_0 has the effect of raising the cell voltage by a constant amount at most currents, and so mimics raising the open circuit voltage (OCV). (See Figure 3.4 above.) The last two points in the above list explain the discrepancy between theoretical OCV and actual OCV noted in Section 2.5.4 in the previous chapter.

3.4.4   Summary of activation overvoltage

In low and medium temperature fuel cells activation overvoltage is the most important irreversibility and cause of voltage drop, and occurs mainly at the cathode. Activation overvoltage at *both* electrodes is important in cells using fuels other than hydrogen, such as methanol. At higher temperatures and pressures the activation overvoltage becomes less important.

Whether the voltage drop is significant at both electrodes, or just the cathode, the size of the voltage drop is related to the current density i by the equation:-

$$V = A \ln\left(\frac{i}{b}\right)$$

where A and b are constants depending on the electrode and cell conditions. This equation is only valid for $i > b$.

3.5   Fuel Crossover and Internal Currents

Although the electrolyte of a fuel cell will have been chosen for its ion conducting properties, it will always be able to support very small amounts of electron conduction. The situation is akin to minority carrier conduction in semiconductors. Probably more important in a practical fuel cell is that some hydrogen will diffuse from the anode through the electrolyte to the cathode. Here, because of the catalyst, it will react directly with the oxygen, producing no current from the cell. This small amount of wasted fuel that migrates through the electrolyte is known as 'fuel crossover'.

These effects – fuel crossover and internal currents - are essentially equivalent. The crossing over of one hydrogen molecule from anode to cathode, where it reacts, wasting 2 electrons, amounts to exactly the same as 2 electrons crossing from anode to cathode internally, rather than as an external current. Furthermore, if the major loss in the cell is the transfer of electrons at the cathode interface, which is the case for hydrogen fuel cells, then the effect of both these phenomena on the cell voltage is also the same.

Although internal currents and fuel crossover are essentially equivalent, and the fuel crossover is probably more important, the effect of these two phenomena on the cell voltage is easier to understand if we just consider the internal current. We, as it were, assign the fuel crossover as "equivalent to" an internal current. This is done in the explanation that follows.

The flow of fuel and electrons will be small, typically the equivalent of only a few mA.cm^{-2}. In terms of energy loss this irreversibility is not very important. However, in low temperature cells it does cause a very noticeable voltage drop at open circuit. Users of fuel cells can readily accept that the working voltage of a cell will be less than the theoretical "no loss" reversible voltage. However, at open circuit, when no work is being done, surely it should be the same! With low temperature cells, such as PEM cells, if operating on air at ambient pressure, the voltage will usually be at least 0.3 volts less than the ~1.2 volts reversible voltage that might be expected.

If, as in the last section, we suppose that we have a fuel cell which only has losses caused by the "activation overvoltage" on the cathode, then the voltage will be as in equation 3.3:-

$$V = E \ - \ A \ \ln\left(\frac{i}{i_o}\right)$$

For the case in point, a PEM fuel cell using air, at normal pressure, at about 30 °C, reasonable values for the constants in this equation are:-

$$E \ = \ 1.2 \ \text{Volts} \qquad A = 0.06 \ \text{Volts} \qquad \text{and} \qquad i_0 = 0.04 \ \text{mA}$$

If we draw up a table of the values of V at low values of current density, we get the following values given in Table 3.2 overleaf.

Table 3.2 Cell voltages at low current densities

Current density mA.cm^{-2}	Voltage volts
0	1.2
0.25	1.05
0.5	1.01
1.0	0.97
2.0	0.92
3.0	0.90
4.0	0.88
5.0	0.87
6.0	0.86
7.0	0.85
8.0	0.84
9.0	0.83

If the internal current density is 2.0 mA.cm^{-2}, then the open circuit voltage will drop to 0.92 volts

Now, because of the internal current density, the cell current density is *not zero*, even if the cell is open circuit. So, for example, if the internal current density is 2 mA.cm^{-2} then the open circuit would be 0.92 volts, nearly 0.3 volts (or 25%) less than the theoretical OCV. This large deviation from the reversible voltage is caused by the very steep initial fall in voltage that we can see in the curves of Figure 3.4. The steepness of the curve also explains another observation about low temperature fuel cells, which is that the OCV is highly variable. The graphs and Table 3.2 tell us that a small change in fuel crossover and/or internal current, caused for example by a change in humidity of the electrolyte, can cause a large change in OCV.

The equivalence of the fuel crossover and the internal currents on the open circuits is an approximation, but is quite a fair one in the case of hydrogen fuel cells where the cathode activation overvoltage dominates. However, the term 'mixed potential' is often used to describe the situation that arises with fuel crossover.

The fuel crossover and internal current are obviously not easy to measure - an ammeter cannot be inserted in the circuit! One way of measuring it is to measure the consumption of reactant gases at open circuit. For single cells and small stacks the very low gas usage rates cannot be measured using normal gas flow meters, and it will normally have to be done using bubble counting, gas syringes, or similar. For example, a small PEM cell of area 10 cm^2 might have an open circuit hydrogen consumption of 0.0034 cm^3.sec^{-1}, at normal temperature and pressure (Author's measurement of a commercial cell.) We know, from Avagadro's law that at STP the volume of one mole of any gas is 2.43 × 10^4 cm^3. So the gas usage is 1.40 × 10^{-7} moles per second.

In Appendix 2 it is shown in equation A2.6 that the rate of hydrogen fuel usage in a single cell (n=1) is related to current by the formula:-

$$Gas\ usage = \frac{I}{2F}\ moles\ s^{-1}$$

so:-

$$I = Gas\ usage \times 2F$$

So in this case the losses correspond to a current I of $1.40 \times 10^{-7} \times 2 \times 9.65 \times 10^{4} = 27$ mA. The cell area is 10 cm^2, so this corresponds to a current density of 2.7 mA.cm^{-2}. This current density gives the total of the current density equivalent of fuel lost because of fuel crossover and the actual internal current density.

If i_n is the value of this internal current density, then the equation for cell voltage that we have been using, equation 3.4, can be refined to :-

$$V = E \quad - \quad A \ \ln\left(\frac{i + i_n}{i_o}\right) \qquad\qquad [3.6]$$

Using typical values for a low temperature cell, viz. $E = 1.2$ Volts, $A = 0.06$ Volts , $i_o = 0.04$ mA.cm^{-2}, and $i_n = 3$ mA.cm^{-2}, we get a graph of voltage against current density that looks like Figure 3.5 below.

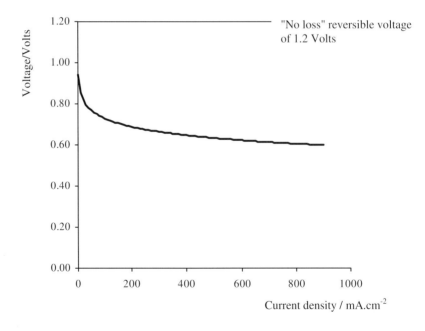

Figure 3.5 Graph showing the fuel cell voltage modelled using activation and fuel crossover/internal current losses only.

The reader will observe that we are already getting a curve that is quite similar to Figure 3.1. The importance of this internal current is much less in the case of higher temperature cells, because the exchange current density i_o is so much higher, and so the initial fall in voltage is not nearly so marked.

So, to sum up, the internal current and/or diffusion of hydrogen through the electrolyte of a fuel cell does not usually have a great importance in terms of operating efficiency. However, in the case of low temperature cells, it has a very marked effect on the open circuit voltage.

3.6   Ohmic Losses

The losses due to the electrical resistance of the electrodes, and the resistance to the flow of ions in the electrolyte, are the simplest to understand and to model. The size of the voltage drop is simply proportional to current, i.e.:-

$$V = I R$$

In most fuel cells the resistance is mainly caused by the electrolyte, though the cell interconnects or bipolar plates (see Section 1.3) can also be important,

To be consistent with the other equations for voltage loss the equation should be expressed in terms of current density. To do this we need to bring in the idea of the resistance corresponding to $1cm^2$ of the cell, for which we use the symbol r. (This quantity is called the "area specific resistance" or ASR.) The equation for the voltage drop then becomes:-

$$V = i r \qquad\qquad [3.7]$$

where i is, as usual, the current density. If i is given in $mA.cm^{-2}$ then the area specific resistance, r, should be given in $k\Omega.cm^2$.

Using the methods described below in Section 3.10 it is possible to distinguish this particular irreversibility from the others. Using such techniques it is possible to show that this "ohmic" voltage loss is important in all types of cell, and especially important in the case of the solid oxide fuel cell (SOFC). Three ways of reducing the internal resistance of the cell are:-

*   The use of electrodes with the highest possible conductivity.

*   Good design and use of appropriate materials for the bipolar plates or cell interconnects. This issue has already been addressed in Section 1.3.

*   Making the electrolyte as thin as possible. However, this is often difficult as the electrolyte sometimes needs to be fairly thick as it is the support onto which the electrodes are built, or it needs to be wide enough to allow a circulating flow of electrolyte. In any case, it must certainly be thick enough to prevent any shorting of one electrode to another through the electrolyte, which implies a certain level of physical robustness.

3.7   Mass Transport or Concentration Losses

If the oxygen at the cathode of a fuel cell is supplied in the form of air, then it is self-evident that during fuel cell operation there will be a slight reduction in the concentration of the oxygen in the region of the electrode, as the oxygen is extracted. The extent of this change in concentration will depend on the current being taken from the fuel cell, and on physical factors relating to how well the air around the cathode can circulate, and how quickly the oxygen can be replenished. This change in concentration will cause a reduction in the partial pressure of the oxygen.

Similarly, if the anode of a fuel cell is supplied with hydrogen, then there will be a slight drop in pressure if the hydrogen is consumed as a result of a current being drawn from the cell. This reduction in pressure results from the fact that there will be a flow of hydrogen down the supply ducts and tubes, and this flow will result in a pressure drop due to their fluid resistance. This reduction in pressure will depend on the electric current from the cell (and hence H_2 consumption) and the physical characteristics of the hydrogen supply system.

Whether we are talking about a reduction in pressure, or a reduction in partial pressure, the same principles apply. In both cases the reduction depends on the electric current and the physical characteristics of the system. The effect of this reduction in pressure or partial pressure can be seen by revisiting equation 2.8. This gives the change in open circuit voltage caused by a change in pressure of the reactants. In equation 2.10 we saw that the change in voltage caused by a change in hydrogen pressure only is:-

$$\Delta V = \frac{RT}{2F} \ln\left(\frac{P_2}{P_1}\right)$$

Now, the change in pressure caused by the use of the fuel gas can be estimated as follows. We postulate a limiting current density i_l at which the fuel is used up at a rate equal to its maximum supply speed. The current density cannot rise above this value, because the fuel gas cannot be supplied at a greater rate. At this current density the pressure will have just reached zero. If P_1 is the pressure when the current density is zero, and we assume that the pressure falls linearly down to zero at the current density i_l then the pressure P_2 at any current density i is given by the formula:-

$$P_2 = P_1\left(1 - \frac{i}{i_l}\right)$$

If we substitute this into equation 2.10 (given above) we obtain:-

$$\Delta V = \frac{RT}{2F} \ln\left(1 - \frac{i}{i_l}\right) \qquad\qquad [3.8]$$

This gives us the voltage change due to the mass transport losses. We have to be careful with signs here, equation 2.10 and 3.8 are written in terms of a voltage *gain*, and the term inside the brackets is always less than 1. So if we want an equation for voltage *drop* we should write it as:-

$$V = - \frac{RT}{2F} \ln\left(1 - \frac{i}{i_l}\right)$$

Now the term that in this case is $\frac{RT}{2F}$ will be different for different reactants, as should be evident from equation 2.8. For example, for oxygen it will be $\frac{RT}{4F}$. In general we may say that the concentration or mass transport losses are given by the equation:-

$$V = - B \ln\left(1 - \frac{i}{i_l}\right) \qquad [3.9]$$

where B is a constant that depends on the fuel cell and its operating state. If this type of loss at just one electrode were the only cause of voltage drop, then the fuel cell operating voltage would be given by the equation:-

$$V = E + B \ln\left(1 - \frac{i}{i_l}\right)$$

If we put in appropriate values for the constants, i.e. $E = 1.2$ Volts, $B = 0.016$ Volts and $i_1 = 1000$ mA, we obtain the graph shown in Figure 3.6.

The extreme non-linearity of this function, when using the theoretical value of B, is clear. In practice the "cut-in" of the mass transport losses is not so sudden, and we might speculate a higher value of B, as in the second curve in Figure 3.6. This is because changes in concentration and partial pressure also affect the activation losses. (Because the exchange current density i_0 is reduced, and the important effect of this has already been discussed in Section 3.4.) This can be modelled by making B rather larger than the theoretical value. However, it is clear that the once the current density reaches the limiting value at one electrode, the voltage falls rapidly to zero, whatever the limiting current density at the other electrode. This means that equation 3.9 gives quite a good representation of the mass transport voltage drop in the whole cell, not just at one electrode, provided that the constant B is larger than $\frac{RT}{2F}$, the theoretical value.

The mass transport or concentration overvoltage is particularly important in cases where the hydrogen is supplied from some kind of reformer, as this might have difficulty increasing the rate of supply of hydrogen quickly to respond to demand. Another important case is at the air cathode, if the air supply is not well circulated. A particular problem is that the nitrogen that is left behind after the oxygen is consumed can cause a mass transport problem at high currents - it effectively blocks the oxygen supply. Note that we used the Nernst equation to derive the formula for these "mass transport" or "concentration" voltage losses, which is why some workers call this type of voltage drop "Nernstian".

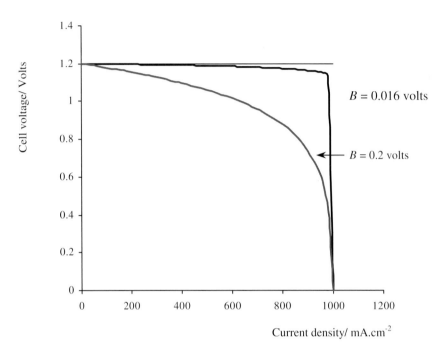

Figure 3.6 The effect of the voltage drop due to mass transport losses at one electrode, for two different values of the constant B.

3.8   Combining the Irreversibilities

It is useful to construct an equation that brings together all these irreversibilities. We can do so and arrive at the following equation for the operating voltage of a fuel cell at a current density density i.

$$V = E - (i+i_n)r - A\ln\left(\frac{i+i_n}{i_0}\right) + B\ln\left(1 - \frac{i+i_n}{i_l}\right) \qquad [3.10]$$

In this equation:-

E is the reversible open circuit voltage given by equation 3.1
i_n is the internal and fuel crossover equivalent current density described in Section 3.5
A is the slope of the Tafel line as described in Section 3.4.2
i_0 is either the exchange current density at the cathode if the cathodic overvoltage is much greater than the anodic, or it is a function of both exchange current densities as given in equation 3.5

B is the constant in the mass transfer overvoltage equation as discussed in Section 3.7

i_l is the limiting current density at the electrode which has the lowest limiting current density, as discussed in Section 3.7

r is the area specific resistance, as described in Section 3.6.

Example values of the constants are given in the table below, for two different types of fuel cell.

Table 3.3 Example constants for equation 3.10

Constant	Low temp., e.g. PEMFC	High temp., e.g. SOFC
E / volts	1.2	1.0
i_n / mA.cm^{-2}	2	2
r / kΩ.cm^2	30×10^{-6}	300×10^{-6}
i_0 / mA.cm^{-2}	0.067	300
A / volts	0.06	0.03
B / volts	0.05	0.08
i_l / mA.cm^{-2}	900	900

It is possible to model this equation using a spreadsheet (such as EXCEL), or graphics calculator. It must be borne in mind that there may be problems at low current densities, as the third term in equation 3.10 is only valid when $(i + i_n) \rangle i_o$. Also, the equation is not valid when the limiting current density is exceeded, i.e. $(i + i_n) \rangle i_l$. This is left as an exercise for the reader. Graphs very like those in Figures 3.1 and 3.2 will be obtained.

The approach we have taken in considering these different losses has been fairly rigorous and mathematical, suitable for an initial understanding of the issues. It is of course possible to be much more rigorous. For such a rigorous theoretical approach, yet soundly based upon a the performance of a real fuel cell stack, the reader is referred to Amphlett et al. (1995).

3.9   The Charge Double Layer

The "charge double layer" is a complex and interesting electrode phenomenon, and whole books have been written on the topic (e.g. Bockris, 1975). However, a much briefer account will suffice in this context. The charge double layer is important in understanding the dynamic electrical behaviour of fuel cells.

Whenever two different materials are in contact there is a build-up of charge on the surfaces or a charge transfer from one to the other. For example, in semiconductor materials, there is a diffusion of "holes" and electrons across junctions between N type and P type materials. This forms a "charge double layer" at the junction, of electrons in the P type region and "holes" in the N type, which has a strong impact on the behaviour of semiconductor devices. In electrochemical systems the charge double layer forms in part due to diffusion effects, as in semiconductors, and also because of the reactions between the electrons in the electrodes and the ions in the electrolyte, and also as a result of applied voltages. For example, the situation of Figure 3.7 below might arise at the cathode of an acid electrolyte fuel cell. Electrons will collect at the surface of the electrode, and H^+ ions will be attracted to the surface of the electrolyte. These electrons and ions, together with the O_2 supplied by to the cathode, will take part in the cathode reaction:-

$$O_2 + 4e^- + 4H^+ \rightarrow 2H_2O$$

The probability of the reaction taking place obviously depends on the density of the charges, electrons and H^+ ions, on the electrode and electrolyte surfaces. The more the charge, the greater the current. However, any collection of charge, such as these electrons and H^+ ions at the electrode/electrolyte interface, will generate an electrical voltage. The voltage in this case is the "activation overvoltage" we have been considering in Section 3.4. So the charge double layer gives an explanation of why the activation overvoltage occurs. It shows that a charge double layer needs to be present for a reaction to occur, that more charge is needed if the current is higher, and so the overvoltage is higher if the current is greater. We can also see that the catalytic effect of the electrode is important, as an effective catalyst will also increase the probability of a reaction - so that a higher current can flow without such a build-up of charge.

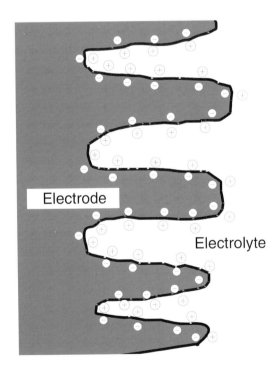

Figure 3.7 The charge double layer at the surface of a fuel cell cathode

The layer of charge on or near the electrode/electrolyte interface is a store of electrical charge and energy, and as such behaves much like an electrical capacitor. If the current changes it will take some time for this charge (and its associated voltage) to dissipate (if the current reduces) or build-up (if there is a current increase). So, the activation overvoltage does not immediately follow the current in the way that the ohmic voltage drop does. The result is that if the current suddenly changes the operating voltage shows an immediate change due to the internal resistance, but moves fairly slowly to its final equilibrium value. One way of modelling this is by using an equivalent circuit, with the charge double layer represented by an electrical capacitor. The capacitance of a capacitor is given by the formula:-

$$C = \varepsilon \frac{A}{d}$$

where ε is the electrical permitivity, A is the surface area and d is the separation of the plates. In this case A is the real surface area of the electrode, which is several thousand times greater than its length $\times$ width. Also d, the separation, is very small, typically only a few nanometres. The result is that, in some fuel cells, the capacitance will be of the order of a few Farads, which is high in terms of capacitance values. (In electrical circuits, a 1 μF capacitor is on the large size of average.) The connection between this capacitance, the charge stored in it, and the resulting activation overvoltage, leads to an equivalent circuit as shown below.

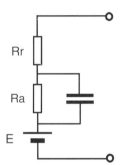

Figure 3.8 Equivalent circuit model of fuel cell

The resistor Rr models the ohmic losses. A change in current gives an immediate change in the voltage drop across this resistor. The resistor Ra models the activation overvoltage, and the capacitor "smoothes" any voltage drop across this resistor. If we were to include the concentration overvoltage this would be incorporated into this resistor too.

Generally speaking, the effect of this capacitance resulting from the charge double layer gives the fuel cell a "good" dynamic performance, in that the voltage moves gently and smoothly to a new value in response to a change in current demand. It also permits a simple and effective way to distinguish between the main types of voltage drop, and hence analyse the performance of a fuel cell, which is described in the next section.

3.10   Distinguishing the Different Irreversibilities

At various points in this chapter it has been asserted that "such and such an overvoltage is important in so and so conditions". For example, it has been said that in SOFC the ohmic voltage drop is more important than activation losses. What is the evidence for these claims?

Some of the evidence is derived from experiments using specialised electrochemical test equipment such as half-cells, which it is beyond the scope of this book to describe. (For such information, see for example Greef et al. (1985).) A method that is fairly straightforward to understand is that of *electrical impedance spectroscopy*. A variable frequency alternating current is driven through the cell, the voltage is measured and the impedance calculated. At higher frequencies the capacitors in the circuits will have less impedance. By plotting graphs of impedance against frequency it is possible to find the values of the equivalent circuit of Figure 3.8. It is sometimes even possible to distinguish between the losses at the cathode and the anode, and certainly between mass transport and activation type losses. Wagner (1998) gives a particularly good example of this type of experiment as applied to fuel cells. However, because the capacitances are large, and the impedances small, special signal generators and measurement systems are needed. Frequencies as low as 10 mHz may be used, so the experiments are often rather slow.

The *current interrupt technique* is an alternative that can be used to give accurate quantitative results (e.g. Lee, 1998), but is also used to give quick qualitative indications. It can be performed using standard low cost electronic equipment. Suppose a cell is providing a current at which the concentration (or mass transport) overvoltage is negligible. The voltage drop will in this case be caused by the ohmic losses and the activation overvoltage. Suppose now that the current is suddenly cut off. The charge double layer will take some time to disperse, and so will the associated overvoltage. However, the ohmic losses will *immediately* reduce to zero. We would therefore expect the voltage to change as in Figure 3.9 below if the load was suddenly disconnected from the cell.

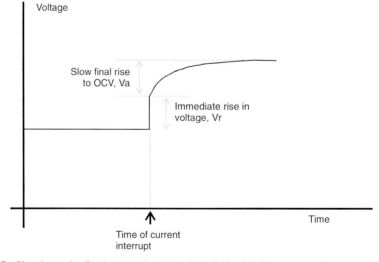

Figure 3.9 Sketch graph of voltage against time for a fuel cell after a current interrupt

The simple circuit needed to perform this current interrupt test is shown in Figure 3.10 below. The switch is closed, and the load resistor adjusted until the desired test current is flowing. The storage oscilloscope is set to a suitable timebase, and the load current is then switched off. The oscilloscope triggering will need to be set so that the oscilloscope moves into 'hold' mode - though with some cells the system is so slow this can be done by hand. The two voltages Vr and Va are then read off the screen. Although the method is simple, when obtaining quantitative results care must be taken, as it is possible to overestimate Vr by missing the point where the vertical transition ends. The oscilloscope timebase setting needed will vary for different fuel cell types, depending on the capacitance, as is done in the three example interrupt tests overleaf. These issues are addressed, for example, by Büchi et al. (1995).[2]

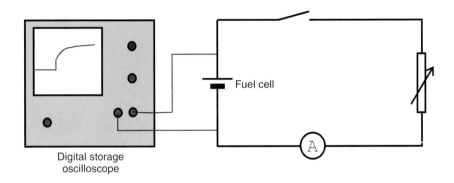

Figure 3.10   Simple circuit for performing a current interrupt test.

The current interrupt test is particularly easy to perform with single cells and small fuel cell stacks. With larger cells the switching of the higher currents can be problematic. Current interrupts and electrical impedance spectroscopy give us two powerful methods of finding the causes of fuel cell irreversibilities, and both methods are widely used.

Typical results from three current interrupt tests are shown in Figures 3.11, 3.12 and 3.13 overleaf. These three examples are shown because of the clear *qualitative* indication they give of the importance of the different types of voltage drop we have been describing. Because oscilloscopes do not show vertical lines, the appearance is slightly different from Figure 3.9, as there is no vertical line corresponding to Vr. The tests were done on three different types of fuel cell, a PEM hydrogen fuel cell, a direct methanol fuel cell, and a solid oxide fuel cell. In each case the *total* voltage drop was about the same, though the current density certainly was not.

---

[2] This paper also outlines an interesting variation on the current interrupt test, in which a pulse of current is applied to the cell.

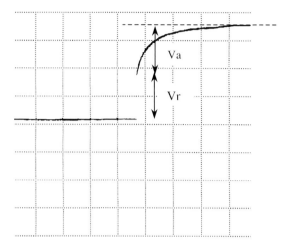

Figure 3.11. Current interrupt test for a low temperature, ambient pressure, hydrogen fuel cell. The ohmic and activation voltage drops are similar. (Time scale 0.2 sec/div, $i=100$ mA.cm^{-2})

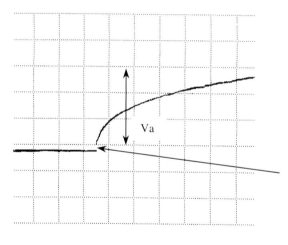

Figure 3.12 Current interrupt test for a direct methanol fuel cell. There is a large activation overvoltage at **both** electrodes. As a result the activation overvoltage is much greater than the ohmic, which is barely discernable. (Time scale 2 sec/div. $i=10$ mA.cm^{-2})

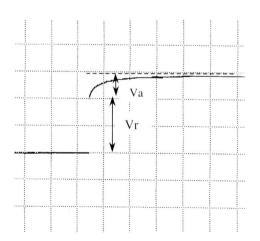

Figure 3.13 Current interrupt test for a small solid oxide fuel cell working at about 700 °C. The large immediate rise in voltage shows that most of the voltage drop is caused by ohmic losses. (Time scale 0.02 sec/div. $i=100$ mA.cm$^{-2)}$

These three examples give a good summary of the causes of voltage losses in fuel cells. Concentration or mass transport losses are important only at higher currents, and in a well designed system, with good fuel and oxygen supply, they should be very small at rated currents. In low temperature hydrogen fuel cells the activation overvoltage (at the cathode) is important, especially at low currents, but the ohmic losses play an important part too, and the activation and ohmic loses are similar (Figure 3.11). In fuel cells using fuels such as methanol then there is a considerable activation overvoltage at *both* the anode and cathode, and so the activation overvoltage dominates at all times (Figure 3.12). On the other hand, in higher temperature cells the activation overvoltage becomes much less important, and ohmic losses are the main problem (Figure 3.13).

We now have a sufficient understanding of the principles of fuel cell operation, and in the following chapters we look much more closely at the practical details of different types of fuel cell systems.

References

Amphlett J.C., Baument R.M., Mann R.E. Peppley B.A., Roberge P.R. Harris T.J. (1995) "Performance modelling of the Ballard Mark V solid polymer electrolyte fuel cell" *Journal of the Electrochemical Society*, Vol. 142, No 1. pp1 - 15

Appleby A.J & Foulkes F.R. (1993), *A Fuel Cell Handbook*, 2nd Ed. Kreiger Publishing Co., p.22

Bloom H., Cutman F. (Eds.) (1981) *Electrochemistry*, Plenum Press, p.121

Bokins J.O'M., Conway B.E., Yeager E.,(Eds.), (1975) *Comprehensive Treatment of Electrochemistry,* Vol.1., Plenum Press,

Büchi F.N., Marek A., Schere G.G. (1995)"In-situ membrane resistance measurements in polymer electrolyte fuel cells by fast auxiliary current pulses" *Journal of the Electrochemical Society*, Vol. 142, No. 6, pp1895 - 1901

Davies C.W., (1967) *Electrochemistry*, Newnes, p.188

Greef R., Peat R., Peter L.M., Pletcher D., Robinson J. (1985) *Instrumental Methods in Electrochemistry*, Ellis Horwood/John Wiley & Sons

Lee C.G., Nakano H., Nishina T., Uchida I., Kuroe S. (1998) "Characterisation of a 100 cm^2 class molten carbonate fuel cell with current interruption", *Journal of the Electrochemical Society*, Vol. 145, No. 8, pp2747-2751

McDougall A., (1976) *Fuel Cells*, Macmillan, pp37-41

Wagner N., Schnarnburger N., Mueller B., Lang M. (1998) "Electrochemical impedance spectra of solid-oxide fuel cells and polymer membrane fuel cells", *Electrochimica Acta*, Vol.43, No.24, pp3785-379

4

Proton Exchange Membrane Fuel Cells

4.1 Overview

The proton exchange membrane fuel cell (PEMFC), also called the "solid polymer fuel cell" (SPFC), was first developed by General Electric in the USA in the 1960's for use by NASA on their first manned space vehicles.

The electrolyte is an ion conduction polymer, described in more detail in Section 4.2 below. Onto each side is bonded a catalysed porous electrode. The anode-electrolyte-cathode assembly is thus one item, and is very thin, as shown in Figure 4.1 overleaf. These "membrane electrode assemblies" (or MEAs) are connected in series, usually using bipolar plates as in Figure 1.8.

The mobile ion in the polymers used is an H^+ ion or proton, so the basic operation of the cell is essentially the same as for the acid electrolyte fuel cell, as shown in Figure 1.3.

The polymer electrolytes work at low temperature, which brings the further advantage that a PEMFC can start quickly. The thinness of the MEAs means that compact fuel cells can be made. Further advantages are that there are no corrosive fluid hazards, and that the cell can work in any orientation. This means that the PEMFC is particularly suitable for use in vehicles and in portable applications.

The early versions of the PEMFC, as used in the NASA Gemini spacecraft, had a lifetime of only about 500 hours, but that was sufficient for those limited early missions. (Warshay, 1990). The development program continued with the incorporation of a new polymer membrane in 1967 called Nafion, a registered trademark of Dupont. This type of membrane, outlined in Section 4.2, became standard for the PEMFC, as it still is today.

However, the problem of water management in the electrolyte, which we consider in some detail in Section 4.4 below, was judged too difficult to manage reliably, and for the Apollo vehicles NASA selected the "rival" alkali fuel cell. General Electric also chose not to pursue commercial development of the PEMFC, probably because the costs were seen as higher than other fuel cells, such as the phosphoric acid fuel cell, then being developed. At the time catalyst technology was such that 28mg of platinum were needed for each cm^2 of electrode – compared to $0.2mg.cm^{-2}$ or less now.

The development of PEM cells went more or less into abeyance in the 1970's and early 1980's. However, in the latter half of the 1980's and early 1990's there was a renaissance of interest in this type of cell (Prater, 1990). A good deal of the credit for this must go to

Ballard Power Systems of Vancouver, Canada and to the Los Alamos National Laboratory in the USA[1].

The developments over recent years have brought the current densities up to around 1 $A.cm^{-2}$ or more, while at the same time reducing the use of platinum by a factor of over 100. These improvements have led to huge reduction in cost per kW of power, and much improved power density, as can be seen from the fuel cell stacks in Figure 4.1 below.

Figure 4.1 Four PEM fuel cell stacks illustrating developments through the 1990s. The left hand stack, the 1989 model, has a power density of 100 $W.L^{-1}$. The right hand, 1996 model is 1.1 $kW.L^{-1}$. (By kind permission of Ballard Power systems.)

PEMFCs are being actively developed for use in cars and buses, as well as for a very wide range of portable applications, and also for combined heat and power systems. A sign of the dominance of this type of cell is that they are again the preferred option for NASA, and the new Space Shuttle Orbiter will use PEM cells (Warshay et al., 1997). It could be argued that PEMFCs exceed all other electrical energy generating technologies in the breadth of scope of their possible applications. They are a possible power source at a few watts for powering mobile phones and other electronic equipment such as computers, right through to a few kW for boats and domestic systems, to tens of kW for cars, to hundreds of kW for buses and industrial CHP systems.

Within this huge range of applications two aspects of PEM fuel cells are more or less similar. These are:

- the electrolyte used, which is described in Section 4.2
- the electrode structure and catalyst. This we cover in Section 4.3

---

[1] The story of the Ballard fuel cell company is told in Koppel (1999)

However, other important aspects of fuel cell design vary greatly depending on the application and the outlook of the designer. The most important of these are:

- water management – a vital topic for PEMFCs dealt with in Section 4.4
- the method of cooling the fuel cell, which we discuss in Section 4.5
- the method of connecting cells in series. The bipolar plate designs vary greatly, and some fuel cells use altogether different methods. We discuss these in Section 4.6
- the question of what pressure to operate the fuel cell, which we consider in Section 4.7
- the reactants used is also an important issue – pure hydrogen is not the only possible fuel, and oxygen can be used instead of air. This is briefly discussed in Section 4.8

Finally, in Section 4.9 we look at some example PEM fuel cell systems. There are, of course, other important questions, such as "where does the hydrogen come from?" but these questions are large, and apply to all fuel cell types, and so are dealt with in separate chapters.

4.2   How the Polymer Electrolyte Works

The different companies producing polymer electrolyte membranes have their own special tricks, mostly proprietary. However, a common theme is the use of sulphonated fluoro-polymers, usually fluoroethylene. The most well known and well established of these is Nafion (® Dupont), which has been developed through several variants since the 1960's. This material is still the electrolyte against which others are judged, and is in a sense an "industry standard". Other polymer electrolytes function in a similar way.

The construction of the electrolyte material is as follows. The starting point is the basic, simplest to understand, man-made polymer - polyethylene. Based on ethylene, its molecular structure is shown in Figure 4.2 below.

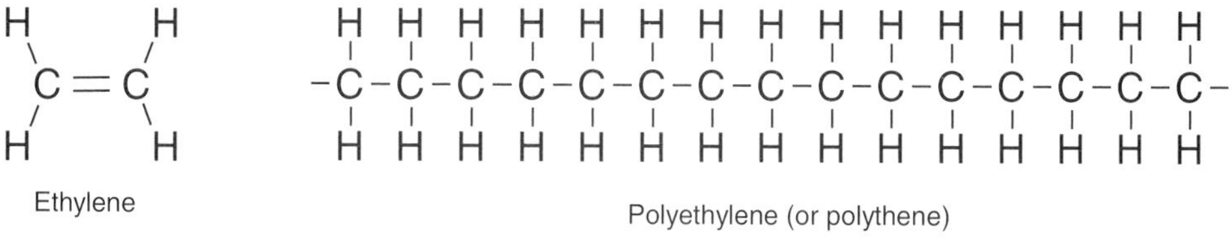

Ethylene   Polyethylene (or polythene)

Figure 4.2 Structure of polyethylene

This basic polymer is modified by substituting fluorine for the hydrogen. This process is applied to many other compounds, and is called 'perfluorination'. The 'mer' is called tetrafluoroethylene[2]. The   modified   polymer,   shown   in   Figure   4.3,   is polytetrafluoroethylene, or PTFE. It is also sold as Teflon, the registered trademark of ICI. This remarkable material has been very important in the development of fuel cells. The strong bonds between the fluorine and the carbon make it resistant to chemical attack and durable. Another important property is that it is strongly hydrophobic, and so it is used in fuel cell electrodes to drive the product water out of the electrode, and thus prevent

[2] 'Tetra' indicates that all four hydrogens in each ethylene group have been replaced by fluorine.

flooding. It is used in this way in phosphoric acid and alkali fuel cells, as well as PEMFCs. (The same property gives it a host of uses in outdoor clothing and footwear.)

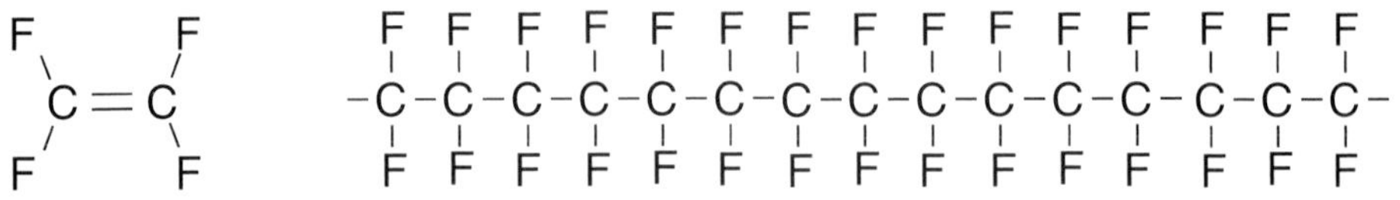

Tetrafluoroethylene   Polytetrafluoroethylene (PTFE)

Figure 4.3 Structure of PTFE

However, to make an electrolyte, a further stage is needed. The basic PTFE polymer is "sulphonated" – a side chain is added, ending with sulphonic acid HSO_3. Sulphonation of complex molecules is a widely used technique in chemical processing. It is used, for example in the manufacture of detergent. One possible side chain structure is shown in Figure 4.4 – the details vary for different types of Nafion, and with different manufacturers of these membranes. The methods of creating and adding the side chains is proprietary, though one modern method is discussed in Kiefer et al., (1999).

The HSO_3 group added is ionically bonded, and so the end of the side chain is actually an SO_3^- ion. For this reason the resulting structure is called an "ionomer". The result of the presence of these SO_3^- and H^+ ions is that there is a strong mutual attraction between the + and – ions from each molecule. The result is that the side chain molecules tend to "cluster" within the overall structure of the material. Now, a key property of sulphonic acid is that it is highly hydrophyllic – it attracts water. (This is why it is used in detergent, it makes one end of the molecule mix readily with water, while the other end attaches to the dirt.) In Nafion, this means we are creating hydrophyllic regions within a generally hydrophobic substance, which is bound to create interesting results!

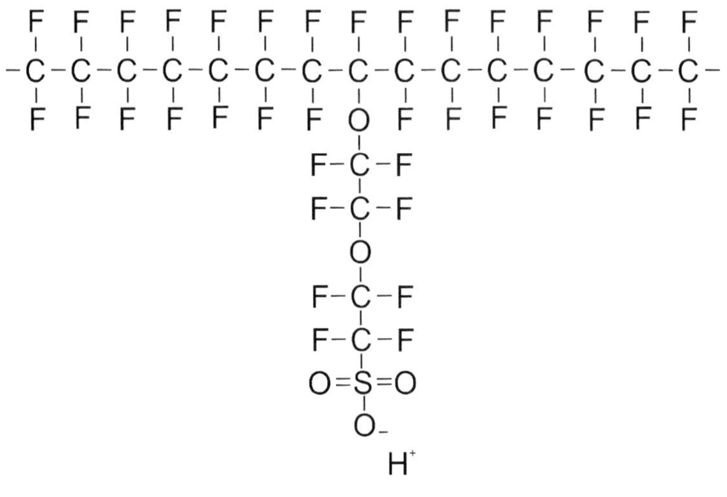

Figure 4.4 Example structure of a sulphonated fluoroethylene, (also called 'perfluorosulphonic acid PFTE copolymer')

The hydrophyllic regions around the clusters of sulphonated side chains can lead to the absorption of large quantities of water, increasing the dry weight of the material by up to 50%. Within these hydrated regions the H⁺ ions are relatively weakly attracted to the SO_3^- group, and are able to move. This creates what is essentially a dilute acid. The resulting material has different phases, dilute acid regions within a tough and strong hydrophobic structure. This "micro-phase separated morphology" is illustrated in Figure 4.5 opposite. Although the hydrated regions are somewhat separate, it is still possible for the H⁺ ions to move through the supporting long molecule structure. However, it is easy to see that for this to happen the hydrated regions must be as large as possible.

From the point of view of fuel cell use, the main features of Nafion and other fluorosulphonate ionomers are that:

- they are highly chemically resistant
- they are mechanical strong, and so can be made into very thin films, down to 50 μm
- they are acidic
- they can absorb large quantities of water
- if they are well hydrated, then H⁺ ions can move quite freely within the material – they are good proton conductors.

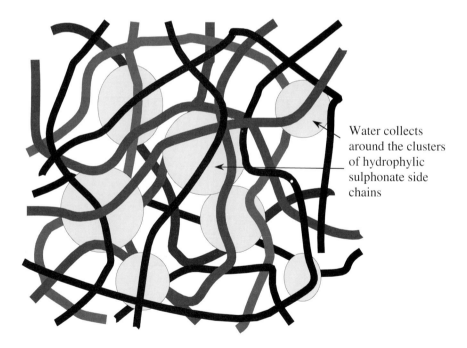

Water collects around the clusters of hydrophylic sulphonate side chains

Figure 4.5 showing the structure of Nafion type membrane materials. Long chain molecules containing hydrated regions around the sulphonated side chains.

4.3   Electrodes and Electrode Structure

The best catalyst for both the anode and the cathode is platinum. In the early days of PEMFC development this catalyst was used at the rate of 28 mg of platinum per cm^2. This high rate of usage led to the myth, still widely held, that platinum is a major factor in the cost of PEMFC. In recent years the usage has been reduced to around 0.2 $mg.cm^{-2}$, yet with power increasing. At such 'loadings' the basic raw material cost of the platinum metal in a 1 kW PEMFC would be about $10 – a small proportion of the total cost.

The basic structure of the electrode in different designs of PEMFC is similar, though of course details vary. The anodes and the cathodes are essentially the same too, for indeed in many PEMFCs they are identical.

The platinum catalyst is formed into very small particles on the surface of somewhat larger particles of finely divided carbon powders. A carbon based power XC72 (® Cabot) is widely used. The result, in somewhat idealised form, is shown below in Figure 4.6. A real picture of this type of supported catalyst is shown in Figure 1.6 on page 6. The platinum is highly divided and spread out, so that a very high proportion of the surface area will be in contact with the reactants.

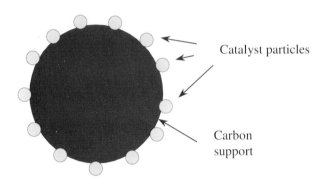

Catalyst particles

Carbon
support

Figure 4.6 The structure, idealised, of a carbon supported
platinum catalyst.

For the next stage one of two alternative routes are used, though the end result is essentially the same in both cases.

In the *separate electrode method* the carbon supported catalyst is fixed, using proprietary techniques, to a porous and conductive material such as carbon cloth or carbon paper. PTFE will often be added also, because it is hydrophobic and so will expel the product water to the surface where it can evaporate. As well as providing the basic mechanical structure for the electrode, the carbon paper or cloth also diffuses the gas onto the catalyst and so is often called the 'gas diffusion layer'. An electrode is then fixed to each side of a piece of polymer electrolyte membrane. A fairly standard procedure for doing this is described in several papers (e.g. Lee et al., 1998). First the electrolyte membrane is cleaned by immersing in boiling 3% hydrogen peroxide in water for one hour, and then in boiling sulphuric acid for the same time, to ensure as full protonation of the sulphonate group as possible. The membrane is then rinsed in boiling de-ionised water for one hour to remove any remaining acid. The electrodes are then put onto the electrolyte membrane, and the assembly hot pressed at 140° C at high pressure for three minutes. The result is a complete membrane electrode assembly, or MEA.

The alternative method involves *building the electrode directly onto the electrolyte*. The platinum on carbon catalyst is fixed directly to the electrolyte, thus manufacturing the electrode directly onto the membrane, rather than separately. The catalyst, which will often (but not always) be mixed with hydrophobic PTFE, is applied to the electrolyte membrane using rolling methods (e.g. Bever et al., 1998), or spraying (e.g. Giorgi et al., 1998) or an adapted printing process (Ralph et al., 1997). With the exception of the first paper describing this idea (Wilson & Gottesfield, 1992), the literature generally gives very little detail of the method used, usually referring to "proprietary techniques". Once the catalyst is fixed to the membrane, a gas diffusion layer must be applied. This will be carbon cloth or paper, as is used for the separate electrodes. 'Gas diffusion layer' is a slightly misleading name for this part of the electrode, as it does much more than diffuse the gas. It also forms an electrical connection between the carbon supported catalyst and

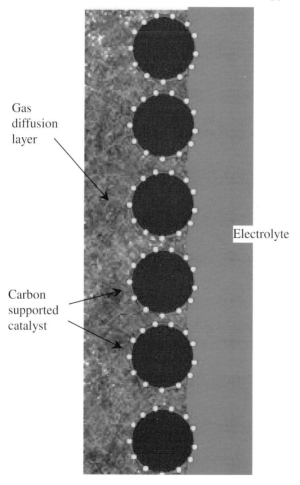

Gas
diffusion
layer

Electrolyte

Carbon
supported
catalyst

Figure 4.7 Simplified and idealised structure of a PEM fuel cell electrode

the bipolar plate, or other current collector. In addition it carries the product water away from the electrolyte surface, and also forms a protective layer over the very thin layer of catalyst. This gas diffusion layer may or may not be an integral part of the membrane electrode assembly.

Whichever of these two methods is chosen, the result is a structure as shown, in idealised form, in Figure 4.7. The carbon supported catalyst particles are joined to the electrolyte on one side, and the gas diffusion (+ current collecting, water removing, physical support) layer on the other. The hydrophobic PTFE that is needed to remove water from the catalyst is not shown explicitly, but will almost always be present.

Two further points need to be discussed. The **first** relates to the impregnation of the electrode with electrolyte material. In Figure 4.8 below a portion of the catalyst / electrode region is shown enlarged. It can be seen that the electrolyte material spreads out over the catalyst. It does not cover the catalyst, but makes a direct connection between catalyst and electrolyte. This increases the performance of the membrane electrode assembly markedly (e.g. Lee et al., 1998), as only the catalyst that is in contact with both the membrane electrolyte and the reaction gas is active. This light covering of the catalyst with the electrolyte is achieved by brushing the electrode with a solubilised form of the electrolyte. In the case of the "separate electrode" method, this is done before the electrode is hot pressed onto the membrane. In the case of the integral membrane / electrode method it is done before the gas diffusion layer is added.

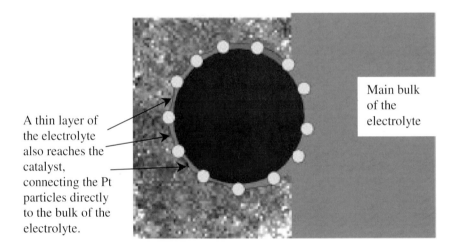

A thin layer of the electrolyte also reaches the catalyst, connecting the Pt particles directly to the bulk of the electrolyte.

Main bulk of the electrolyte

Figure 4.8 Enlargement of part of Figure 4.7, showing that the electrolyte reaches out to the catalyst particles

The **second** relates to the selection of the gas diffusion layer. We have seen that this is generally either a carbon paper or carbon cloth material. Carbon paper (e.g. Toray® paper is a widely used brand) is chosen when it is required to make the cell as thin as possible in compact designs. Carbon cloths are thicker, and so will absorb a little more water, and also simplify mechanical assembly, since they will fill small gaps and irregularities in bipolar plate manufacture and assembly. On the other hand they will slightly expand out into the gas diffusion channels on the bipolar plates, which may be significant if these have been made very shallow, as will be the case in a very thin design. In the case of the very low power small PEMFCs to be discussed in Sections 4.4 and 4.5 below, even thicker materials, such as the carbon felt may be used, as a comparatively wide cell separation is needed for air circulation.

So, we have the heart of our proton exchange membrane fuel cell, which is the membrane electrode assembly. No matter how this is made, or which company made it, the MEA will look similar, work in essentially the same way, and require similar care in use. However, the way in which working fuel cells stacks are made around these MEAs varies enormously. In the following four Sections we look at some of these different approaches, taking as our themes the ways in which the major engineering problems of the PEMFC are solved.

Figure 4.9 Samples of membrane electrode assemble (MEA) mounted on gaskets for placing in a stack. The upper sample has some gas diffusion material attached.

4.4   Water Management in the PEMFC

4.4.1   Overview of the problem

It will be clear from the description of a proton exchange membrane given in Section 4.2 that there must be sufficient water content in the polymer electrolyte, otherwise the conductivity will decrease. However, there must not be so much water that the electrodes, which are bonded to the electrolyte, flood, blocking the pores in the electrodes or gas diffusion layer. A balance is therefore needed, which takes care to achieve.

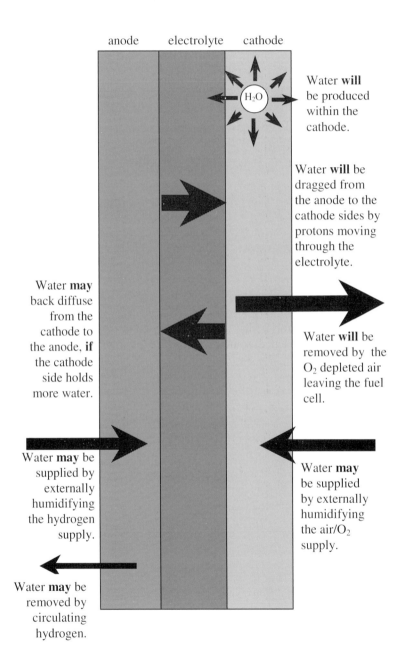

anode   electrolyte   cathode

Water **will** be produced within the cathode.

Water **will** be dragged from the anode to the cathode sides by protons moving through the electrolyte.

Water **may** back diffuse from the cathode to the anode, **if** the cathode side holds more water.

Water **will** be removed by the O_2 depleted air leaving the fuel cell.

Water **may** be supplied by externally humidifying the hydrogen supply.

Water **may** be supplied by externally humidifying the air/O_2 supply.

Water **may** be removed by circulating hydrogen.

Figure 4.10 Showing the different water movements to, within, and from the electrolyte of a PEM fuel cell.

In the PEMFC water forms at the cathode – revisit Figure 1.3 on page 3 if you are not sure why. In an ideal world this water would keep the electrolyte at the correct level of hydration. Air would be blown over the cathode, and as well as supplying the necessary oxygen it would dry out any excess water. Because the membrane electrolyte is so thin, water would diffuse from the cathode side to the anode, and throughout the whole electrolyte a suitable state of hydration would be achieved without any special difficulty. This happy situation can sometimes be achieved, but needs good engineering design to bring to pass.

There are several complications. One is that during the operation of the cell the H^+ ions moving from the anode to the cathode (see Figure 1.3) pull water molecules with them. This process is sometimes called "electro-osmotic drag". Typically between 1 and 2.5 water molecules are "dragged" for each proton (Zawodzinski et al., 1993). This means that, especially at high current densities, the anode side of the electrolyte can become dried out – even if the cathode is well hydrated. Another major problem is the drying effect of air at high temperatures. We will show this quantitatively below in Section 4.4.2, but suffice to say at this stage that at temperatures of over about 60°C the air will *always* dry out the electrodes faster than water is produced by the H_2/O_2 reaction. One common way to solve these problems is to humidify the air, the hydrogen, or both, before they enter the fuel cell. This may seem bizarre, as it effectively adds by-product to the inputs to the process, and there can't be many other processes where this is done! However, we will see that this is sometimes needed, and greatly improves the fuel cell performance.

Yet another complication is that the water balance in the electrolyte must be correct throughout the cell. In practice, some parts may be just right, others too dry, and others flooded. An obvious example of this can be seen if we think about the air as it passes through the fuel cell. It may enter the cell quite dry, but by the time it has passed over some of the electrodes it may be about right. However, by the time it has reached the exit it may be so saturated it cannot dry off any more excess water. Obviously, this is more of a problem when designing larger cells and stacks.

All these different water movements are shown in Figure 4.10 opposite. Fortunately all these water movements are predictable and controllable. Starting from the top of Figure 4.10, the water production and the water drag are both directly proportional to the current. The water evaporation can be predicted with care, using the theory outlined below in Section 4.4.2. The back diffusion of water from cathode to anode depends on the thickness of the electrolyte membrane, and the relative humidity of each side. Finally, if external humidification of the reactant gases is used prior to entry to the fuel cell, this is a process which can of course be controlled.

4.4.2   Air flow and water evaporation

Except for the special case of PEM fuel cells supplied with pure oxygen, it is universally the practice to remove the product water using the air that flows through the cell. The air will also always be fed through the cell at a rate faster than that needed just to supply the necessary oxygen. If it were fed at exactly the 'stoichiometric' rate then there would be very great "concentration losses", as described in the previous chapter, Section 3.7. This is because the exit air would be completely depleted of oxygen. In practice the stoichiometry (λ) will be at least 2. In Appendix 2, Section A2.2, a very useful equation is derived connecting the air flow rate, the power of a fuel cell and the stoichiometry (equation A2.4). Problems arise because the drying effect of air is so non-linear in its relationship to temperature. To understand this we have to consider the precise meaning and quantitative effects of terms such as *relative humidity, water content* and *saturated vapour pressure*.

When considering the effect of oxygen concentration in Section 2.5, the partial pressures of the various gases that make up air were given. (See page 31.) At that point we ignored the fact that air also contains water vapour. We did this because the amount of water vapour in air varies greatly, depending on the temperature, location, weather conditions, and other factors. A straightforward way of measuring and describing the amount of water vapour in air is to give the ratio of water to other gases - the other gases being nitrogen, oxygen, argon, carbon dioxide and others that make up "dry air". This quantity is variously known as the *humidity ratio, absolute humidity*, or *specific humidity* and is defined as:-

$$humidity\ ratio,\ \omega = \frac{m_w}{m_a} \qquad [4.1]$$

where m_w is the mass of water present in the sample of the mixture, and m_a is the mass of dry air. The total mass of the air is $m_w + m_a$.

However, this does not give a very good idea of the drying effect or the "feel" of the air. Warm air, with quite a high water content, can feel very dry, and indeed have a very strong drying effect. On the other hand, cold air, with a low water content, can feel very damp. The reason for this is due to the changes in the *saturated vapour pressure* of the water vapour. The saturated vapour pressure is the *partial pressure* of the water when a mixture of air and liquid water is at equilibrium - the rate of evaporation is equal to the rate of condensation. The air cannot hold any more water vapour, it is "saturated". This is illustrated in Figure 4.11 below.

Air that has no "drying effect", that will not be hold any more water, could reasonably be said to be "fully humidified". This state is achieved when $P_w = P_{sat}$, where P_w is the partial pressure of the water, and P_{sat} is the saturated vapour pressure of the water. We define the *relative humidity* as the ratio of these two pressures:-

$$relative\ humidity,\ \phi = \frac{P_w}{P_{sat}} \qquad [4.2]$$

Typical relative humidities vary from about 0.3 (or 30%) in the ultra-dry conditions of the Sahara desert to about 0.7 (or 70%) in New York on an "average day". Very important for us is the fact that the drying effect of air, or the rate of evaporation of water, is directly proportional to the difference between the water partial pressure P_w and the saturated vapour pressure P_{sat}.

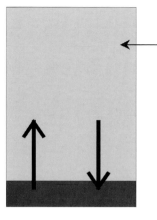

Water/air mixture. The partial pressure of the water vapour is **equal** to the *saturated vapour pressure*. There is thus an equilibrium. The water content of the air does not change, it is "saturated" and can take no more.

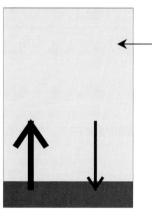

Water/air mixture. The partial pressure of the water vapour is **less than** the *saturated vapour pressure*. The water at the bottom of the vessel will evaporate faster than any condensation. The rate of evaporation of water is directly proportional to the difference between the partial pressure of the water P_w and the saturated vapour pressure P_{sat}.

$$Rate\ of\ evaporation \propto (P_{sat} - P_w)$$

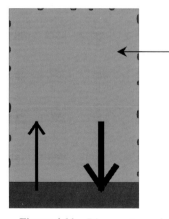

Water/air mixture. The partial pressure of the water vapour is **greater** than the *saturated vapour pressure*. In this case the rate of condensation exceeds the rate of evaporation. The condensation will usually be observed on the walls of the vessel. This time the rate of condensation is proportional to the difference in the two water vapour pressures:-

$$Rate\ of\ condensation \propto (P_w - P_{sat})$$

Figure 4.11   Diagram to explain saturated vapour pressure

The cause of the complication for PEM fuel cells is that the saturated vapour pressure varies with temperature in a highly non-linear way - P_{sat} increases more and more rapidly at higher temperatures. The saturated vapour pressure for a range of temperatures is given in Table 4.1 below.

Table 4.1 The saturated vapour pressure of water at selected temperatures

T, $^\circ$C	Saturated Vapour Pressure/kPa
15	1.705
20	2.338
30	4.246
40	7.383
50	12.35
60	19.94
70	31.19
80	47.39
90	70.13

The result of the rapid rise in P_{sat} is that air that might be only moderately drying, say 70% relative humidity, at ambient temperature, can be fiercely drying when heated to about 60 °C. For example, for air at 20 °C, relative humidity 70%, the pressure of the water vapour in the mixture is:-

$$P_w = 0.70 \times P_{sat} = 0.70 \times 2.338 = 1.64 \quad \text{kPa}$$

If this air is then heated to 60°C, at constant pressure, without adding water, then P_w will not change, and so the new relative humidity will be:-

$$relative\ humidity,\ \phi = \frac{P_w}{P_{sat}} = \frac{1.64}{19.94} = 0.08 = 8\%$$

This is very dry, far more drying than the Sahara desert for example, for which ϕ is typically about 30% [data for the Dakhla Oasis in Daly (1979)]. This would have a catastrophic effect upon polymer electrolyte membranes, which not only rely totally on a high water content, but are also very thin, and so prone to rapid drying out.

Another way of describing the water content that should be mentioned is the *dew point*. This is the temperature to which the air should be cooled to reach saturation. For example, if the partial pressure of the water in a sample of air is 12.35 kPa, then, referring to Table 4.1, the *dew point* would be 50 °C.

To judge whether or not water needs to be added to the incoming gas we need to carefully consider the water produced by the operation of the fuel cells. To calculate the effect of this water on the humidity we need to use the following equations:-

- Equation 4.1 above, giving the humidity ratio.
- Equation A2.10, from Appendix 2, where the rate of production of water in a fuel cell is derived.

$$Rate\ of\ water\ production = 9.34 \times 10^{-8} \times \frac{P_e}{V_c} \quad kg.s^{-1}$$

where P_e is the electrical output power and V_c is the mean voltage of each cell in the fuel cell stack.

- Equation A2.5, from Appendix 2, where the exit air flow rate was shown to be:-

$$Exit\ air\ flowrate = \left(3.57 \times 10^{-7} \times \lambda - 8.29 \times 10^{-8}\right) \times \frac{P_e}{V_c} \quad kg.s^{-1}$$

where λ is the stoichiometry of the air, typically at least 2.

If we ignore the water content of the input air, and we take the water produced and air flowing out during one second, then combining equations 4.1, A2.5 and A2.10 the humidity ratio of the air leaving the cell is:-

$$humidity\ ratio,\ \omega = \frac{m_w}{m_a} = \frac{9.34 \times 10^{-8}}{(3.57 \times 10^{-7} \times \lambda - 8.29 \times 10^{-8})} \qquad [4.3]$$

Note that we are assuming that all the product water is evaporated. If we also, quite reasonably, assume that the water vapour and air are behaving as perfect gases, then, from their molar masses we can say that:-

$$\omega = \frac{18.016}{28.97} \times \frac{P_w}{P_a} = 0.622 \times \frac{P_w}{P_a} \qquad [4.4]$$

However, the total air pressure is the sum of the dry air and water vapour pressures, so:-

$$P_t = P_a + P_w \qquad \therefore\ P_a = P_t - P_w$$

Substituting this into equation 4.4 gives:-

$$\omega = 0.622 \times \frac{P_w}{P_t - P_w} \qquad \therefore P_w = \frac{\omega P_t}{\omega + 0.622}$$

If we substitute equation 4.3 for ω in this, and do some simplification too tedious to include here, we obtain:-

$$P_w = \frac{0.421}{\lambda + 0.188} P_t \qquad [4.5]$$

This shows that the water vapour pressure in the exit air depends, quite simply, on the air stoichiometry and the operating pressure P_t. We should also add to this partial pressure the pressure of the water vapour at the inlet, but at low values of λ the effect will be small. The effect of P_t is important; it shows that we get a higher humidity (P_w), for otherwise similar conditions, if the cell is operated at higher pressure. We will come back to this when we consider the question of fuel cell operating pressure in Section 4.7.

Equation 4.5 can then be used, together with the saturated vapour pressure taken from Table 4.1, to calculate the exit air humidity at different temperatures. Graphs of the humidity at air stoichiometries of 2 and 4 are shown in Figure 4.12 for a cell operating at 100 kPa (1 bar). Some selected figures are also given in Table 4.2 It can be readily seen that for most operating conditions the fuel cell will either be too wet or too dry!

As one would expect, the humidities are lower at greater air flow ($\lambda = 4$). Also, at higher temperatures the relative humidity falls sharply. If the relative humidity of the exit air is much less than 100%, then the effect will be for the cell to dry out, and the PEM to cease working. This may not be obvious, but remember that these figures are calculated assuming that *all* the product water is evaporated. In these circumstances, if the exit air is still less than 100% relative humidity then the fuel cell has given all it has got to give, and yet the air still wants more! It must also be remembered that the conditions at the entry will be *even more drying*. The only way to solve this problem is to humidify one or both of the reactant gases.

On the other hand, a relative humidity of greater than 100% is basically impossible, and the air stream will contain condensed water droplets. This can be dealt with in moderation, with the water blown out by the air. However, if the theoretical humidity is much greater than 100% then the electrodes will become flooded. The result of this two-way constraint is a fairly narrow band of operating conditions. Nevertheless, if we stay below about 60°C, then there will be an air flowrate that will give a suitable humidity, i.e. around 100%. Some of these conditions are shown in Table 4.2 below, and practical ways of achieving the required levels of humidification throughout the cell are discussed in Section 4.4.3 below.

A very important point from Figure 4.12 and Table 4.2 is that *at temperatures of above about 60°C, the relative humidity of the exit air is below or well below 100% at all reasonable values of stoichiometry.* (If λ is less than 2, then the oxygen concentration will be too low for the cells near the exit air flow.) This leads to the important conclusion that ***extra humidification of the reactant gases is essential in PEM fuel cells operating above about 60°C.*** This has been confirmed by the general experience of PEM fuel cell users. (e.g. Büchi and Srinivasan, 1997). The methods for humdifying the gases are discussed in Section 4.4.4 below.

This feature makes for difficulties in choosing the optimum operating temperature for a PEMFC. The higher the temperature, the better the performance, mainly because the cathode overvoltage reduces. However, once over 60°C the humidification problems increase, and the extra weight and cost of the humidification equipment can exceed the savings coming from a smaller and lighter fuel cell.

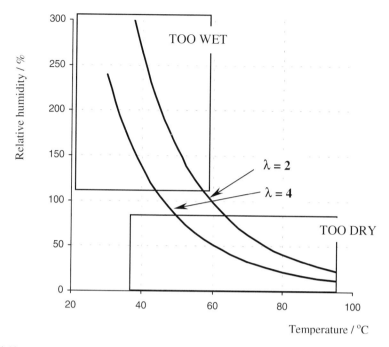

Figure 4.12 A graph of relative humidity vs. temperature for the exit air of a PEM fuel cell at two different stoichiometries. The entry air is assumed to be dry, and the total pressure is 1 bar.

Table 4.2 Exit air relative humidities at selected temperatures and stoichiometries. In this case the inlet air is assumed to be at 20 °C and 70% relative humidity. The gaps are where the relative humidity is absurdly high or low. (Figures are for air pressure of 1.0 bar.)

Temp/°C	$\lambda = 1.5$	$\lambda = 2$	$\lambda = 3$	$\lambda = 6$	$\lambda = 12$	$\lambda = 24$
20					218	145
30				199	120	79
40		282	201	114	69	46
50	215	169	120	68	41	27
60	133	104	74	42		
70	85	67	48			
80	56	44	31			
90	38	30				

4.4.3   Running PEM fuel cells without extra humidification

It is obviously desirable to run PEM fuel cells without externally humidifying the reactant gases - it reduces cost, size and complexity.   There are several papers published considering the pros and cons of external humidification, but a good place to start is Büchi and Srinivasan, 1997). Here it is shown that even below 60°C the maximum power from a fuel cell reduces by about 40% if no external humidification is used. However, in small systems, this is a price worth paying. The overall efficiency of the cell need not reduce, it is just that the voltage begins to fall off at a lower current.

The key to running a fuel cell without external humidification is to set the air stoichiometry so that the relative humidity of the exit air is about 100%, *and* to ensure that the cell design is such that the water is balanced within the cell. One way of doing this is described by Büchi and Srinivasan, 1997), and shown in Figure 4.13. The air and hydrogen flows are in opposite directions across the MEA. The water flow from anode to cathode is the same in all parts, as it is the 'electro-osmotic drag', and is directly proportional to the current. The back diffusion from cathode to anode varies, but is compensated for by the gas circulation. Other aids to an even spread of humidity are narrow electrodes, and thicker gas diffusion layers, which hold more water.

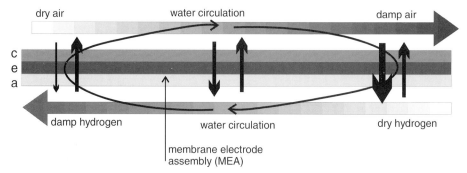

Figure 4.13 Contraflow of reactant gases to spread humidification (Büchi and Srinivisan, 1997).

Referring to Table 4.2, it can be seen that it will always be possible, if the temperature is kept below 60°C, to find a stoichiometry which will give an air humidity of around 100%. At lower temperatures this will be about $\lambda = 24$. This may seem a wildly extravagant airflow, but in practice it is not. Let us take the example of the small type of fuel cell to which this sort of technique applies. The smallest fan from the catalogue of a major manufacturer has a flowrate of 4.7 cfm (= ft^3. min^{-1}, the manufacturer is in the USA!). If we assume that only 2.0 cfm of this air actually flows over the cathodes, and that the stack operates with an average cell voltage of 0.6 volts, then, using equation A2.4 (adapted for cfm), we find that this airflow is suitable for a fuel cell of power:-

$$P_e = \frac{0.6 \times 2.0}{3.57 \times 10^{-7} \times 24 \times 1795} = 78 \quad \text{Watts}$$

The electrical power of the fan in question is just 0.7 watts, so the air circulation is supplied using less than 1% of the fuel cell power, at a stoichiometry of 24. This is a very modest parasitic power loss.

4.4.4   External humidification

Although we have seen that small fuel cells can be operated without additional or external humidification, in larger cells this is rarely done. Operating temperatures of over 60°C are desirable to reduce losses, especially the activation voltage drops described in Section 3.4. Also, it makes economic sense to operate the fuel cell at maximum possible power density, even if the extra weight volume, cost and complexity of the humidification system is taken into account. With larger cells, all these are proportionally less important.

In laboratory test systems the reactant gases of fuel cells are humidified by simply bubbling them through water, whose temperature is controlled. However, this will rarely be practical in the field. Systems used are usually taken or adapted from the humidifiers used in air conditioning systems. The easiest to control is the direct injection of water as a spray. This has the further advantage that it will cool the gas, which will be necessary if it has been compressed (see Section 4.6 below) or if the fuel gas has been formed by reforming some other fuel, as in Chapter 7. A problem special to fuel cells is that the water used for humidification must be as pure as practical – impurities in the water hydrating the proton exchange membrane will adversely affect its performance. This is usually dealt with by cooling the air leaving the cell and condensing out the water, thus giving what is essentially distilled water.

Although the use of equipment adapted from air conditioning equipment could be described as 'mainstream', there are other methods that have been described for humidification that are special to fuel cells. The term "external" humidification is not entirely appropriate for some of these. 'Super-humidification' might be more appropriate, as extra water is applied or generated, but sometimes within the fuel cell. It could well be that some of these could become standard practice as the industry grows. Three of these other methods are outlined below.

1. A definitely external system of great simplicity has been demonstrated by engineers at the Paul Scherrer Institute in Switzerland. The principle is shown in Figure 4.14 opposite. The warm damp air leaving the cell passes over one side of a membrane, where it is cooled. Some of the water condenses on the membrane. The liquid water passes through the membrane and is evaporated by the drier gas going into the cell on the other side. Such a humidifier unit can be seen on the top of the fuel cell system shown in Figure 4.29. In this case, a 2.0 kW fuel cell, only the air is externally humidified. The MEA is particularly thin, and this permits the anode side to be sufficiently hydrated by back diffusion.

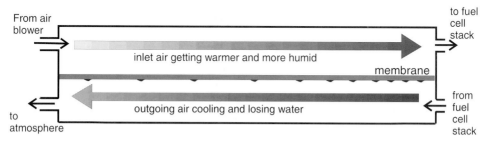

Figure 4.14 Humidification of reactant air using exit air, as demonstrated by the Paul Scherrer Institute (1999). See also Figure 4.29

2. Another approach is to directly inject liquid water into the fuel cell. Normally this would lead to the electrode flooding, and the cell ceasing to work. However, the technique is combined with a bipolar plate and "flow field" design that forces the reactant gases to blow the water through the cell and over the entire electrode. This is well described by Wood et al., 1998), who have coined the term "interdigitated flow

field" for the method. The principle is shown in Figure 4.15 overleaf. The "flow field" is the name of the channel cut in the bipolar plate that is the route for the reactant gas. The flow field shown in Figure 4.15 is like a maze with no exit! The gas is forced under the bipolar plate and into the electrode, driving the water with it. If the flow field is well designed this will happen all over the electrode. Good results are reported for this method, though the reactant gases must be driven at pressure through the cell, and the energy used to do this is not clear. Neither is the wear and tear on the electrodes and the effect on long term performance.

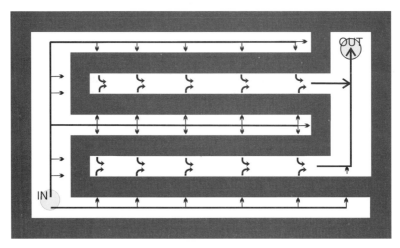

TOP VIEW

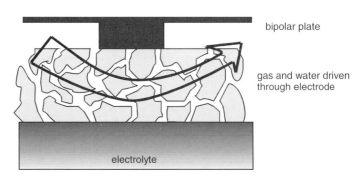

SIDE VIEW, ENLARGED

Figure 4.15 Diagrams to show the principle of humidification using interdigitated flow fields, after Wood et al. (1998).

3. Watanabe (1996) has described a system called "self-humidification", where the electrolyte is modified, not only to retain water, but also to *produce* water. Retention is increased by impregnating the electrolyte with particles of silica (SiO_2) and titania (TiO_2), which are hygroscopic materials. Nanocrystals of platinum are also impregnated into the electrolyte, which is made particularly thin. Some hydrogen and oxygen diffuse through the electrode and, because of the catalytic effect of the platinum, react, producing water. This of course uses up valuable hydrogen gas, but it is claimed that the improved performance of the electrolyte justifies this parasitic fuel loss.

The actual method used and extent of humidification will vary depending on the size of the cell, the operating pressure (to be considered below in Section 4.7), the balance sought between optimum performance and simplicity, the fuel source, among other considerations. In any case the problem will need careful thought at the design stage. In larger systems the extra humidification of the reactant gases will need to be actively controlled.

4.5   PEM Fuel Cell Cooling and Air Supply

4.5.1   Cooling using the cathode air supply

PEM fuel cells are, of course, not 100% efficient. In converting the hydrogen energy into electricity, efficiencies are normally about 50%. This means that a fuel cell of power X watts will also have to dispose of about X watts of heat. More precisely, it is shown in Appendix 2, Section A2.6, that the heat produced by a fuel cell, if the product water is evaporated within the cell, is

$$Heating \ \ rate = P_e \left(\frac{1.25}{V_c} - 1 \right) \ \ \text{Watts}$$

The way this heat is removed depends greatly on the size of the fuel cell. With fuel cells below 100 watts it is possible to use purely convected air to cool the cell and provide sufficient airflow to evaporate the water, without recourse to any fan. This is done with a fairly open cell construction with a cell spacing of between 5 and 10mm per cell (e.g. Daugherty, 1999). The fact that damp air is less dense than dry air, (perhaps counter-intuitive, but true!) aids the circulation process. However, for a more compact fuel cell small fans can be used to blow the reactant and cooling air through the cell, though a large proportion of the heat will still be lost through natural convection and radiation. As was shown with the numerical example at the end of Section 4.4.3, this does not impose a large parasitic power loss on the system – only about 1% if the fan is well chosen and the air ducting sensibly designed.

However, when the power of the fuel cell rises, and a lower proportion of the heat is lost by convection and radiation from and around the external surfaces of the cell, problems begin to arise. In practice, this simplest of all methods of cooling a fuel cell can only be used for systems of power up to about 200 Watts.

4.5.2   Separate reactant and cooling air

The need to separate the reactant air and the cooling air for anything but the smallest of PEM fuel cells can be shown by working through a specific example where the reactant gas and the cooling gas **are** combined.

Suppose a fuel cell of power P_e watts is operating at 50 °C. The average voltage of each cell in the stack is 0.6 volts - a very typical figure. Let us suppose that cooling air enters the cell at 20 °C, and leaves at 50 °C. (In practice the temperature change will probably not be so great, but let us take the best possible case for the present.) Let us also assume that only 40% of the heat generated by the fuel cell is removed by the air - the rest is radiated or naturally convected from the outer surfaces.

In Appendix 2, Section A2.6, it is shown that the heat generated by a fuel cell, if the water exits as a vapour, is:-

$$Heating \ rate = P_e \left(\frac{1.25}{V_c} - 1 \right) \ Watts$$

Just 40% of this heat is removed by air, of specific heat capacity c_p flowing at a rate of $\dot{m}$ kg.s^{-1}, and subject to a temperature change ΔT. So we can say that:-

$$0.4 \times P_e \left(\frac{1.25}{V_c} - 1 \right) = \dot{m} \, c_p \, \Delta T$$

Substituting known values, i.e. c_p = 1004 J.kg^{-1}.K^{-1}, ΔT = 30 K, and V_c = 0.6 volts, and rearranging, we obtain the following equation for the cooling air flowrate:-

$$\dot{m} = 1.4 \times 10^{-5} \times P_e \quad kg.s^{-1}$$

In Appendix 2, Section A2.2, it is shown that the reactant air flowrate is:-

$$\dot{m} = 3.57 \times 10^{-7} \times \lambda \times \frac{P_e}{V_c} \quad kg.s^{-1}$$

If the reactant air and the cooling air are one and the same, then these two quantities are equal, and so the two equations can be equated. Cancelling P_e, substituting $V_c = 0.6$ volts, and solving for λ we obtain:-

$$\lambda = \frac{14 \times 0.6}{0.357} \approx 24$$

A glance at Table 4.2 above will show that at 50°C this gives an exit air humidity of 27%. This is dryer than the Sahara desert! The figures in Table 4.2 assumed that the entry air had a humidity of 70%, so the relative humidity is *decreasing* as the air goes through the cell, and so the proton exchange membrane will be quickly drying out.

Note that if the assumptions made at the beginning of this section are made more realistic, that is more heat having to be taken out by the air and less of a temperature change, then the situation becomes even worse. The only way to reduce λ, which should be somewhere between 3 and 6 at 50 °C in order to stop the cell drying out, is to reduce the air flowing over the electrodes, and have a separate cooling system. This point comes when more than about 25% of the heat generated has to be removed by a cooling fluid. In practice, this seems to be in the region of about 200 or 300 watts. Fuel cells larger than this will generally need a separate reactant air supply and cooling system. This will mean two air blowers or pumps, but there is no alternative!

The usual way of cooling cells in the range from about 250 to 2500 watts is to make extra channels in the bipolar plates through which cooling air can be blown, as shown in

Figure 4.16 below. Alternatively, separate cooling plates can be added, through which air is blown. Using separate cooling air in this way works for cells between about 250 watts and a few kW, but for larger cells this becomes impractical, and water cooling is preferred.

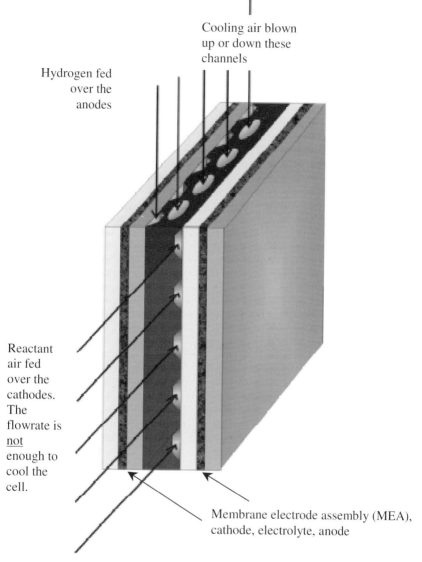

Figure 4.16 Two MEAs and one bipolar plate modified for separate reactant and cooling air.

4.5.3   *Water cooling of PEM fuel cells*

The issues of when to change from air cooling to water cooling are much the same for fuel cells as they are for other engines, such as internal combustion engines. Essentially air cooling is simpler, but it becomes harder and harder to ensure that the whole fuel cell is cooled to a similar temperature as it gets larger. Also, the air channels make the fuel cell stack larger than it need be - one kg of water can be pumped through a much smaller channel than one kg of air, and the cooling effect of water is much greater.

With fuel cells the need to water cool is perhaps greater than with a petrol engine, as the performance is more affected by variation in temperature. On the other hand, the water cooling of a petrol engine also serves as a sound proofing, which is why it is sometimes used, say, on motorcycles when air cooling might otherwise suffice. On balance then, we would expect the "changeover" from water cooling to air cooling to come at around the same sort of power levels, that is at a few kW. PEM fuel cells above 10 kW will generally be water cooled, those below 2 kW air cooled, with the decision for cells in between being a matter of judgement.

One factor that will certainly influence the decision of whether or not to water cool will be the question of "what is to be done with the heat". If it is to be just lost to the atmosphere, then the bias will be towards air cooling. On the other hand, if the heat is to be recovered, for example in a small domestic combined heat and power system, then water cooling becomes much more attractive. It is far easier to "move heat about" if the energy is held in hot water than if it is in air.

The method of water cooling a fuel cell is essentially the same as for air in Figure 4.16, except that water is pumped through the cooling channels. In practice cooling channels are not always needed or provided at every bipolar plate. Variations on this theme are mentioned when we look more closely at different fuel cell construction methods in Section 4.6 below.

4.6   PEM Fuel Cell Construction Methods

4.6.1   Introduction

Most PEM fuel cells are constructed along the general lines of multiple cells connected in series with bipolar plates outlined in Section 1.3, and illustrated in Figure 1.8. However, there are many variations in the way the bipolar plate is constructed and the materials they are made from. This is a very important topic, because as we have seen, the MEAs for PEM fuel cells are very thin, and so the bipolar plates actually comprise almost all the volume of the fuel cell stack, and typically about 80% of the mass. (Murphy et al., 1998] We have also mentioned that the platinum usage has been drastically reduced, making this have little impact on the cost. The result is that the bipolar plates are usually a very high proportion of the cost of a PEM fuel cell stack.

Another important issue is that, especially for smaller PEMFC, there are altogether different ways of making the cells, which avoid the need for a bipolar plate. These two approaches are discussed in this section.

4.6.2   Bipolar plates for PEM fuel cells

The bipolar plate has to collect and conduct the current from the anode of one cell to the cathode of the next, while distributing the fuel gas over the surface of the anode, and the oxygen/air over the surface of the cathode. As well as this, it often has to carry a cooling fluid though the stack, and keep all these reactant gases and cooling fluids apart. The distribution of the reactant gases over the electrodes is done using a "flow field" formed into the surface of the plate, which is usually a fairly complex serpentine pattern.

The methods for forming these bipolar plates vary considerably, and the most common or promising are described below.

One of the most well established is by the **machining of graphite sheet**. Graphite is electrically conductive, and reasonably easy to machine. It also has a very low density, less than for any metal that might be considered suitable. A well established method of producing the necessary cooling channels within the stack is to make each bipolar plate in two halves, which are identical. The back of each piece has the cooling channels cut in it, and the front the reactant gas flow field. Two such pieces are put back to back to make a complete bipolar plate. Fuel cell stacks made in this way have been made with competitive power density. However, they have three major disadvantages:-

- The machining of the graphite may be done automatically, but the cutting still takes a long time on an expensive machine.
- Graphite is brittle, and so the resulting cell needs careful handling, and assembly is difficult.
- Finally, graphite is quite porous, and so the plates need to be a few mm thick to keep the reactant gases apart. This means that although the material is low density, the final bipolar plate is not particularly light.

Another method that is reasonably well established, though not widely used, is the use of **carbon-carbon composites.** This process is derived from the procedure used for phosphoric acid fuel cells. A composite part made from carbon and a graphitisable resin is made by injection moulding. The graphitisation process is achieved by heating the part to over 2500 °C. The moulding process may be cheap, but the heating process is not. Furthermore, it must be very precisely controlled, otherwise a high proportion of the resulting plates will be warped or the wrong size (Murphy et al., 1998). Dimensional stability through the heating process also means that they cannot be made less than a few mm thick.

A process that is certainly cheap is the **injection moulding of graphite filled polymer** resins. However, even the best of these have such a poor conductivity that the scope of their application is likely to be highly limited for fuel cells in the foreseeable future.

An alternative to injection moulding which lends itself well to bipolar pate construction is **compression moulding**. The material does not need to be nearly so fluid – essentially a piece of sheet material is squeezed to the right shape. This allows a much greater proportion of carbon to be used in the polymer/carbon mixture, and adequate conductivity can be achieved. The shapes produced may not be so intricate, for example the cooling channels in the plates of Figure 4.16 could not be achieved in one piece. However, these problems can be solved by making the plate in two pieces. Reactant gas flow channels are on the front, cooling air channels on the back. A complete bipolar plate is then made by putting the two pieces back to back.

Metals can also be used to make bipolar plates. These have the advantage that they are good conductors, can be machined easily, and are not porous, and so very thin pieces will serve to keep the reactant gases apart. Their major disadvantage is that they have higher density, and are prone to corrosion - the inside of a PEM fuel cell is quite a corrosive atmosphere. There is water vapour, oxygen and warmth. As well as this, there is sometimes a problem of acid leaching out of the MEAs - excess sulphonic acid from the sulphonation process mentioned in Section 4.2.

Metal pieces can be machined in just the same way as graphite sheet. This overcomes the problem of brittleness, and the greater impermeability means that they can be made thinner. However, the time, energy and cost involved in machining the flow fields will be even greater. A better approach is to use **perforated or foamed metal** to make up the flow fields. A particularly successful attempt at this approach has been described by Murphy et al. (1998), and the broad thrust of their method is described here.

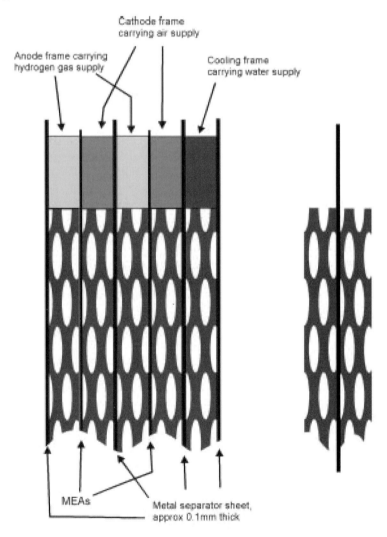

Figure 4.17 Diagram showing bipolar plate construction using metal foam, after Murphy et al. (1998)

The flow field is made from metal which is "foamed" so that it has a sponge like structure, with many small voids within it, and the voids taking up more than 50% of the bulk of the material. The material can then be sawn into thin sheets. The result could be compared to a slice of bread, only it is about 1 or 2 mm thick. An alternative way of making a functionally similar material is to take a solid sheet and introduce the voids by slitting and stretching. The metal needs to be chosen so that it will remain strong and light in this form, and titanium was chosen in the study in question. Titanium is not a particularly good conductor as metals go, but has a resistivity 30 times lower than graphite. Titanium can also be made sufficiently corrosion resistant by coating with a titanium nitride protective finish. This would give an electrically conductive, precious metal-free coating that can be applied inexpensively on a large scale.

The bipolar plate is then formed from two such pieces of foamed metal, with a thin layer of solid metal between, as in the right hand part of Figure 4.17. The voids form the path for the gas to diffuse through. Because the foam has been sliced, like a loaf of bread, much of the surface consists of holes, and it is through these that the gas reaches the surface of the electrodes. Reactant gas is fed into the metal foam at the edges. Plastic holders round the edges of the both hold the metal in place, and provide for the connection to the gas supply. The stack is formed up from layers of metal sheet, metal foam, MEA, metal foam, metal sheet, metal foam, MEA, metal foam,...... etc. as in the left hand part of Figure 4.17.

The mechanism for cooling the cell stack can be provided using essentially the same technology. This time one sheet of the metal foam is placed between two sheets of the solid (but thin) metal sheets. Water is passed through the foamed metal, carrying away the heat. One such heat exchanger is shown in Figure 4.17.

The great advantage of this type of approach is that it uses readily available materials - metal foam sheet is made for other applications - to make fuel cell components that are thin, light, highly conductive, but good at separating gases. Furthermore, the only manufacturing process involved are cutting and moulding. There are many ways in which this basic idea can be adapted. For example, the metal foam can be replaced by two pieces of perforated metal sheet, placed on top of each other, but "out of phase", so that the holes in one do not exactly match the holes in the other. This would allow the reactant gas to flow through the resulting medium.

4.6.3   Other Topologies

Fuel cell stack construction using bipolar plates gives very good electrical connection between one cell and the next. However, it does have the problem that there are many joints and potential problems of reactant gas and cooling fluid leaks. The reactant gas supply to each and every anode has to be kept separate from each and every cathode. The entire edge of each anode and cathode is also a potential leak. These problems can be solved, but it takes careful, and thus expensive, manufacture. In cases where a fuel cell is operated at fairly low current densities, it is sometimes helpful to compromise the electrical resistance of the cell interconnects for simpler and cheaper manufacture. The flexibility and ease of handling of the MEAs used in PEM fuel cells allows many different types of construction.

One such way of doing this is shown in Figure 4.18. This shows a system of three cells. The main body of the unit, shown light grey, would normally be made of plastic. A key point is that there is only one chamber containing air, and only one containing hydrogen. The cells are connected in series by connecting the edge of one cathode to the edge of the next anode. This is done using metal passing through the reactant gas separator. For even less chance of leaks this could be done externally, though this would increase the current path.

The potential leaks are now much reduced, as the only key seals are those around the edges of the MEAs, but if they are carefully set into the separator as in Figure 4.18, which is not hard to do, then this will not be any problem. This type of design also reduces the problem of even humidification of the cells, since there is fairly free circulation of reactant gases in the cell. However, it is not a compact design, and is only suitable for low power systems.

Another way of making small PEM fuel cells worthy of note is the use of a cylindrical design, where the electrode is fully exposed to the air. Basically the simple flat structure of Figure 4.18 is rolled around a hydrogen cylinder, with some extra space between the anodes and the cylinder to allow circulation of hydrogen gas. The fuel cell and hydrogen store are thus integrated. The air supply is supplied very simply by natural air circulation around the outside of the fuel cell. The current collector over the top of the air cathode will be fairly rugged; perforated stainless steel being a possible material.

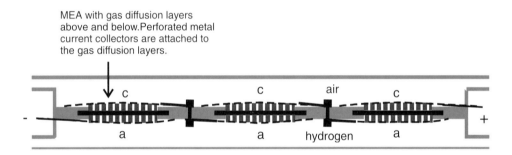

Figure 4.18 A simple method of connecting fuel cells in series, giving far simpler reactant gas supply arrangements

In addition to the problem of higher electrical resistance, a potential problem with these types of fuel cell construction methods is that of cell "imbalance". The individual cells are quite some way apart and they are much more physically separate than in the normal 'bipolar plate type' stack. Suppose one cell becomes rather warmer than the others, this would lead to more rapid water evaporation, which could lead to higher internal resistance, and thus even higher temperatures, leading to more evaporation, higher resistance, ... and so on in a "vicious circle" where one cell could end up dried out and with very high resistance. This would scupper the whole fuel cell, as of course it is a series circuit, and so they all carry the same current. One way of reducing the likelihood of this is to design the system so that they operate at *below* their optimum temperature, so that a small local rise in temperature does not lead to excessive water loss, but this is obviously not ideal. However, it is a further reason why this type of construction is only practical in fairly small systems.

Another fuel cell stack construction method, which is quite close to the 'metal bipolar plates' described in the previous section, has been developed and demonstrated by Advanced Power Sources Ltd of Loughborough in the UK. The construction is as shown in Figure 4.19 below.

Each single cell is constructed from a stainless steel base, MEA, and then a porous metal current collector on top of the cathode. This current collector is made, using patented and proprietary techniques, from sintered stainless steel powder of carefully graded size. The result is a material that is metallic, corrosion resistant, porous, strong, conductive and water retaining. A fuel cell stack is made by placing these rugged and self contained cells one on top of the other, with a simple piece of folded stainless steel connecting the anode of one cell to the cathode of the next. Hydrogen is piped, using thin plastic tubing, to each anode. The open structure of the cell allows for free circulation of air, though this might be fan assisted. There is good thermal (as well as electrical) connection between each cell, so the problem of cell imbalance outlined above is very unlikely.

Consideration of alternative types of cell construction methods are described in the literature, and a good place to start is Heinzel et al. (1998).

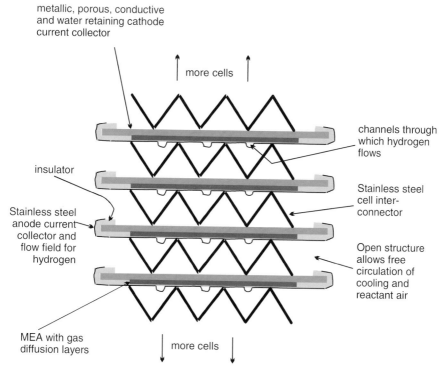

Figure 4.19 PEM fuel cell structure along the lines of that demonstrated by Advanced Power Sources Ltd.

4.7   Operating Pressure

4.7.1   Outline of the problem

Although small PEM fuel cells are operated at normal air pressure, larger fuel cells, of 10kW or more, are sometimes operated at higher pressures. The advantages and disadvantages of operating at higher pressure are complex, and the arguments are not at all clear-cut, there is much to be said on both sides!

The basic issues around operating at higher pressure are the same as for other engines, such as diesel and petrol internal combustion engines, only with these machines the term used is "supercharging" or "turbocharging". Indeed, the technology for achieving the higher pressures is essentially the same. Further on, in Chapter 8, we give a full description of the various types of compressors and turbines that can be used. For a full understanding of this section, the reader is encouraged to look at that chapter now; alternatively the equations derived there can just be accepted at this stage, for a quicker appreciation of the issues.

The purpose of increasing the pressure in an engine is to increase the specific power, to get more power out of the same size engine. Hopefully, the extra cost, size and weight of the compressing equipment will be less than the cost, size and weight of simply getting the extra power by making the engine bigger! It is a fact that most diesel engines are operated at above atmospheric pressure – they are supercharged using a turbocharger. The hot exhaust gas is used to drive a turbine, which drives a compressor, which compresses the inlet air to the engine. The energy used to drive the compressor is thus essentially "free", and the turbocharger units used are mass-produced, compact and highly reliable. In this case the advantages clearly outweigh the disadvantages. However, in the case of the petrol engine, the situation is more complex. The benefits of higher pressure are much less clear cut – increased air intake to the engine can sometimes cause pre-ignition and "knocking". So even when petrol engines are supercharged it is only quite modestly, to about 1.6 bar maximum, compared to up to 3 bar in some diesel engines. In the case of petrol engines then, the situation is much more balanced. Only a fairly small proportion of petrol engines are supercharged. The situation with PEM fuel cells is much more like petrol engines than diesel engines!

The simplest type of pressurised PEM fuel cell is that where the hydrogen gas comes from a high pressure cylinder. Such a system is shown in Figure 4.20 below. Only the air has to be compressed. The hydrogen gas is coming from a pressurised container, and thus its compression "comes free". The hydrogen is fed to the anode in a way called "dead-ended", there is no venting or circulation of the gas - it is just consumed by the cell. The compressor for the air must be driven by an electric motor, which of course uses up some of the valuable electricity generated by the fuel cell. In a worked example in Chapter 8 it is shown that the typical power consumption will be about 20% of the fuel cell power for a 100 kW system. Other systems described in the literature (e.g. Barbir, 1999) report even higher proportions of compressor power consumption. It is also shown in Chapter 8 that the compressed air will need to be cooled before entry to a PEM cell – on internal combustion engines such coolers are also used, and are called "intercoolers".

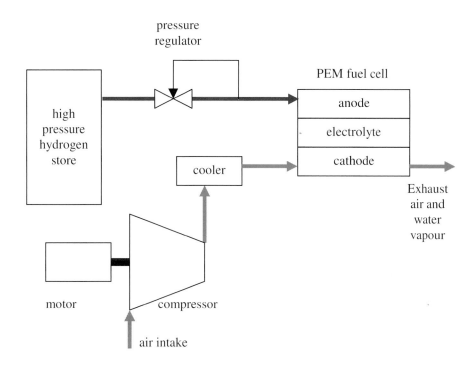

Figure 4.20 Simple motor driven air compressor and PEM fuel cell

When the hydrogen fuel is derived from other hydrocarbons, such as methane, the situation is much more complex. Here the concept of 'balance of plant', mentioned in Chapter 1, comes into play, and the fuel cell becomes apparently "lost" in a mass of other equipment. The methods of reforming the fuel are dealt with in some detail in Chapter 7, but an introductory diagram of one possible method is given above in Figure 4.21. From a point of view of "whether to pressurise or not" the key point is that there is a burner, from which hot gas is given out. This gas can be used to drive a turbine, which can drive the compressor. A motor will still often be needed to start the system, but once in operation it might not be needed at all. Indeed, it is possible to run such a system so that there is an excess of turbine power, and the motor becomes a generator, boosting the electrical power output.

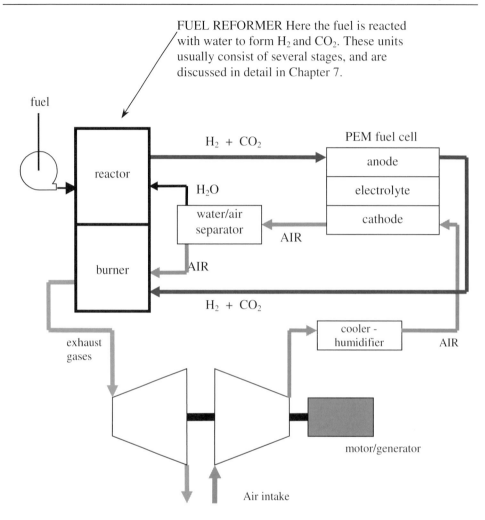

FUEL REFORMER Here the fuel is reacted
with water to form H_2 and CO_2. These units
usually consist of several stages, and are
discussed in detail in Chapter 7.

fuel

reactor

H_2 + CO_2

H_2O

water/air
separator

AIR

burner

AIR

H_2 + CO_2

PEM fuel cell

anode

electrolyte

cathode

cooler -
humidifier

AIR

exhaust
gases

motor/generator

Air intake

Figure 4.21 One possible method of using a turbo-compressor unit to harness the energy of the
exhaust gas from the fuel reformation system to drive the compressor. The compressor supplies
compressed air for the burner that is needed in the fuel reformer. Fuel reformation is covered in
Chapter 7, and turbo-compressors in Chapter 8.

4.7.2   *Simple quantitative cost/benefit analysis of higher operating pressures*

Running a fuel cell at higher pressure will increase the power, but it also involves the
expenditure of power, and there is the cost, weight and space taken up by the compression
equipment. To consider the pros and cons of adding the extra apparatus we have to
consider more *quantitatively* the costs and the benefits of running at higher pressure. This
we do in this section.

The increase in power resulting from operating a PEM fuel cell at higher pressure is mainly the result of the reduction in the cathode activation overvoltage, as discussed in Section 3.4. The increased pressure raises the exchange current density, which has the apparent effect of lifting the open circuit voltage, as shown in Figure 3.4 on page 43. The open circuit voltage really is also raised, because of the change in the Gibbs free energy, as discussed in Section 2.5, and specifically 2.5.4 on page 33. As well as these benefits, there is also sometimes a reduction in the mass transport losses, with the effect that the voltage begins to fall off at a higher current. The effect of raising the pressure on cell voltage can be seen from the graph of voltage against current shown in Figure 4.22. In simple terms, for most values of current, the voltage is raised by a fixed value.

What is not shown on this graph is that this voltage "boost", ΔV, is proportional to the *logarithm* of the pressure rise. This is both an experimental and a theoretical observation. On page 33 we saw that the rise in open circuit voltage due to the change in Gibbs free energy was given by the equation:

$$\Delta V = \frac{RT}{4F} \ln\left(\frac{P_2}{P_1}\right)$$

We also saw in equation 3.4 in Chapter 3 that the activation overvoltage is related to the exchange current by a logarithmic function. So, we can say that, if the pressure is

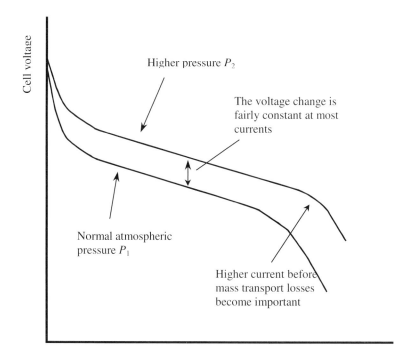

Figure 4.22 The effect of increasing pressure on the voltage/current graph of a typical fuel cell.

increased from P_1 to P_2, then there is an increase or gain in voltage:-

$$\Delta V_{gain} = C \ln\left(\frac{P_2}{P_1}\right) \quad \text{Volts} \qquad [4.6]$$

where C is a constant, whose value depends on how the exchange current density i_o is affected by pressure, and also on R, T and F. Unfortunately it is not clear what value should be used for C. In Hirschenhofer et al. (1995) figures are reported ranging from 0.03 to 0.06 volts. It is certainly known that the benefit depends on the extent of the cell humidification, with more benefit if the cell is not humidified (Büchi and Srinivasan, 1997). Higher values, of up to 0.10 volts can be inferred from some results. However, for each system design, the value for the particular cell in question will have to be found.

For an initial attempt at a cost benefit analysis of pressurisation, the simple system shown in Figure 4.20 is a good place to start. In power terms the cost is straightforward – it is the power needed to drive the compressor. The benefit is the greater electrical power from the fuel cell. As we have seen, for each cell in the stack the increase in voltage ΔV is given equation 4.6 above. To consider the power gain, we suppose a current I Amps and a stack of n cells. The increase in power is then given by :-

$$Power \ gain = C \ln\left(\frac{P_2}{P_1}\right) I n \quad \text{Watts} \qquad [4.7]$$

However, as we have seen, this increase in power is not without cost. In our simple system of Figure 4.20 the power cost is that needed to drive the compressor. We show further on, in Chapter 8, Section 8.4, that an equation can be written for the power consumed by the compressor in terms of the compressor efficiency η_c, the entry temperature of the air T_1, and the pressure ratio P_2/P_1. This power is power that is lost, and so, using equation 8.7, we can say that:-

$$Compressor \ power = power \ lost = c_p \frac{T_1}{\eta_c}\left(\left(\frac{P_2}{P_1}\right)^{\frac{\gamma-1}{\gamma}} - 1\right)\dot{m} \quad \text{Watts}$$

However, this power loss is just the power delivered to the *rotor* of the compressor. This power comes from an electric motor, which has an efficiency of less than 1, and there are also losses in the connecting shaft and the bearings of the compressor rotor. If we express the combined efficiency of the motor and drive system as η_m then the electrical power used will greater than the compressor power by a factor of $1/\eta_m$, and so there will be a loss of *electrical* power given by the equation:-

$$Power \ lost = c_p \frac{T_1}{\eta_m \eta_c}\left(\left(\frac{P_2}{P_1}\right)^{\frac{\gamma-1}{\gamma}} - 1\right)\dot{m} \quad \text{Watts} \qquad [4.8]$$

In this equation $\dot{m}$ is the flowrate of the air, in $kg.s^{-1}$. It is shown in Appendix 2, equation A2.4, that this is connected to the fuel cell electrical power output, the average cell voltage, and the air stoichiometry, by the equation:-

$$\dot{m} = 3.57 \times 10^{-7} \times \lambda \times \frac{P_e}{V_c} \quad kg.s^{-1}$$

Substituting this, and electrical power $P_e = n\,I\,V_c$, and values of c_p and γ for air into equation 4.8 above gives:-

$$\boxed{Power\ lost = 3.58 \times 10^{-4} \times \frac{T_1}{\eta_m\,\eta_c}\left(\left(\frac{P_2}{P_1}\right)^{0.286} - 1\right)\lambda\,I\,n} \quad Watts \qquad [4.9]$$

This term here in equation 4.9 is equivalent to a voltage - compare it with equation 4.7. The term is multiplied by current, and the number of cells, to give a power. It is thus a "loss equivalent" to the voltage *gain* of equation 4.6. It represents the voltage of each cell that is "used up" driving the compressor. We can therefore write an equation that gives us the voltage loss resulting from an increase in pressure from P_1 to P_2 as:-

$$\Delta V_{loss} = 3.58 \times 10^{-4} \times \frac{T_1}{\eta_m\,\eta_c}\left(\left(\frac{P_2}{P_1}\right)^{0.286} - 1\right)\lambda \quad Volts \qquad [4.10]$$

We are now in a position to quantitatively consider whether a pressure increase will improve the net performance of the fuel cell system. We have a simple equation for the increase in voltage, equation 4.6. We also have a straightforward algebraic equation for the voltage loss, or rather voltage "used up", by providing the increased pressure, equation 4.10. What we are now able to do is plot graphs of:-

$$Net\ \Delta V = \Delta V_{gain} - \Delta V_{loss} \quad \text{for different values of } \frac{P_2}{P_1}.$$

However, before we can do this we need to estimate suitable values for the various constants in equations 4.6 and 4.10. In Figure 4.23 this is done for two sets of values, one designated 'optimistic', the other 'realistic'. The values for both these models are given below.

- The voltage gain constant C has already been discussed. 0.06 volts would be a realistic value, though it might be as high as 0.10 volts, which is used as the 'optimistic' value.
- The inlet gas temperature T_1 is taken as 15°C (288 K) for both cases
- The efficiency of the electric drive for the compressor, η_m, is taken as 0.9 for the 'realistic' model, and 0.95 for the 'optimistic'.
- The efficiency of the compressor η_c is taken as 0.75 for the 'optimistic' model, and 0.70 for the 'realistic'.
- The lowest reasonable value for the stoichiometry, λ, is 1.75, and so this is used in the 'optimistic' model. A more realistic value is 2.0.

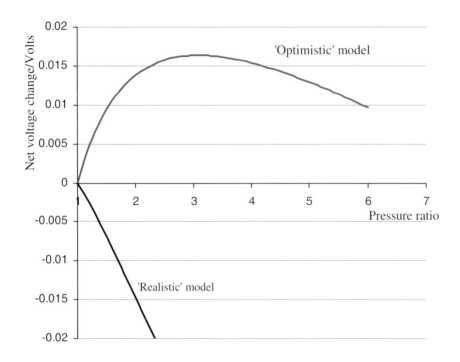

Figure 4.23 Graphs of net voltage change resulting from operating at higher pressure, for two different PEM fuel cell models.

These values are used in equations 4.6 and 4.10 to plot graphs of the net voltage change, as shown in Figure 4.23 above. It can be seen that for the optimistic model, there is a net gain of about 17mV per cell when the pressure is boosted by a ratio of about 3, but the gains diminish at higher pressures. However, for the more 'realistic' model, there is *always* a net loss as a result of the higher pressure! The power gained is always exceeded by the power needed to drive the compressor. We can see why the practice of operating at above atmospheric pressure is by no means universal even with larger PEMFCs. With smaller fuel cells, less than about 5kW, it is very rare, as the compressors are likely to be less efficient and more expensive, as is explained in Section 7 of Chapter 8.

The system we have considered is very simple, yet we have been able to build a model against which to quantitatively judge the issue of the optimum pressure. However, this is only an introduction to the topic of fuel cell system modelling. There are other issues that must be considered, other factors that might make higher operational pressures more or less attractive. These we consider in the next section.

4.7.3   Other factors affecting choice of pressure

Although it is the simplest to quantify, the voltage boost is not the only benefit from operating at higher pressure. Similarly, loss of power to the compressor is not the only loss.

One of the most important gains can be shown by reference to Figure 4.21. In this system there is a fuel reforming system. Such chemical plant usually benefits from higher pressure. In this particular case, as we shall see in Chapter 7, the kinetics of the reaction indicate that a lower pressure would be preferred. However, this is more than compensated for by the reduction in size that is achieved by higher pressure. So, as well as the fuel cell itself, the fuel reformation system also gains from higher pressure.

In the system of Figure 4.21 there is a burner, which is needed to provide heat for the fuel reformation process. The exhaust from this burner can be used by a turbine to drive the compressor - thus the energy for the compression process becomes essentially "free". This is not really true, because, as can be seen by a careful study of the diagram, the energy comes from hydrogen that has passed through the fuel cell, but was not turned into electricity. Typically about 10 to 20% of the hydrogen will be needed for the fuel reformer. However, this burning and heating process is necessary for the fuel reformer, so we are not running the turbine off anything that would not otherwise be burnt. This greatly affects the analysis, as of course we no longer have the electrical power loss associated with running a compressor.

Another factor of great importance in favour of pressurisation is the question of reactant air humidification. Humidification is a great deal easier if it can be combined with air cooling, because there is plenty of energy available to evaporate the water. A temperature rise is always associated with compression of gases. In Chapter 8, where we consider compressors in much more detail, we show in a worked example that a typical temperature rise would be about 150 °C. Air at such a temperature could be very easily humidified, and the process would also cool the air closer to a suitable temperature. However, the main benefit is that less water is needed to achieve the same humidity at higher pressures, and at higher temperatures the difference is particularly great. It can be seen from Table 4.1 that above 80 °C the saturated vapour pressure rises very markedly, which means that the partial pressure of the dry air *falls* very greatly. This means that the humidity ratio increases even more. However, if the pressure is high, then the partial pressure of the air will not be so greatly effected. The example below should make things clear.

Example. Compare two samples of air at 90 °C, one at 100 kPa, the other at 300 kPa. What humidity ratio is needed to achieve 100 % relative humidity for these two samples?

At 100 kPa, P_w = 70 kPa, so P_a = 30 kPa, so from equation 4.4, $\omega = 0.622 \times \dfrac{70}{30} = 1.45$

At 300 kPa, P_w = 70 kPa, so P_a = 230 kPa, so from equation 4.4, $\omega = 0.622 \times \dfrac{70}{230} = 0.19$

The humidity ratio tells us the mass of water that needs to be present in the air to achieve the required humidity. So, we can see that at 300 kPa, the water needed is less by a factor of nearly 8. If we take into account factors such as the water that was in the air before compression, and the product water, we find that the mass of water to be added via humidification reduces by an even greater factor. So, for higher temperature PEM fuel cells, this is a most important consideration.

With larger fuel cells the flow paths for the reactant gases will be quite long, and if size has been minimised, narrow. Thus a certain pressure will be needed to get them through the cell in any case! If this pressure is about 0.3 bar, the designer might then be faced with the problem of choosing an available compressor or blower, and might find that they are not available at that pressure, but only at higher pressures. In which case the choice will have been forced the very practical issue of product availability - air blowers and compressors in the pressure range of around 0.2/0.3 bar are not readily available at the smaller flowrates required by fuel cells.

On the negative side there are the issues of size, weight and cost. However, it must be borne in mind that some sort of air blower for the reactant air would be needed anyway, so it is the *extra* size, weight and cost of higher pressure compressors compared to lower pressure blowers that is the issue. The practical issue of product availability means that this difference will often be quite small for fuel cells of power in the region of tens of kW.

With systems such as that in Figure 4.20 the compressor will often be a screw compressor, as described in Chapter 8. A problem with this type of compressor is that they can be very noisy, certainly compared to the silence of the rest of the system. This is an important problem, since it negates one of the major advantages of fuel cells over other electricity generators. The centrifugal compressors (also described in Chapter 8) associated with the turbo units of Figure 4.21 are somewhat quieter, but nevertheless increased noise will often be an important disadvantage of a pressurised system.

In the discussion thus far we have assumed that air is being used to supply the cathode with oxygen. There are some systems, notably in space applications, where pure oxygen from pressurised cylinders, is used. In these cases the choice of operating pressure will be made on completely different grounds. The issues will be the balance between increased performance and increased weight due to the mechanical strength needed to operate at the higher pressures. The optimum pressure will probably be much higher than for air systems.

In the last section we constructed a simple model to try to find an optimum operating pressure. Real models would have to take into account the issues we have been discussing in this section, most notably the effect of pressure on the fuel reformer, and the effect of the turbine. Such models would be highly non-linear and discontinuous - pumps and blowers are not readily available at all pressure ratios and all flowrates. There will be a huge number of variables, and many different ways the system could be run. For example, the system could be run at higher fuel stoichiometry, giving more heat out of the burner, and more power at the turbine, which could be converted into electricity in the motor/generator unit. Would this give more total electrical power output than a set-up optimised for the fuel cell itself? A careful model would have to be constructed to find out. This can be done using commercially available mathematical modelling computer programs.

In practice, most small systems operate at approximately ambient air. It is the larger systems, that often operate at higher pressure.

4.8   Reactant Composition

4.8.1   Carbon monoxide poisoning

Up to now we have generally assumed that our PEM fuel cells have been running on pure hydrogen gas as the fuel, and air as the oxidant. In small systems this will frequently be the case. However, in larger systems the hydrogen fuel will frequently come from some kind of fuel reforming system. These are discussed in much more detail in Chapter 7, and have been mentioned briefly back in Section 4.7.1, and in Figure 4.21. These fuel reformation systems nearly always involve a reaction producing carbon monoxide, such as the reaction between methane and steam:-

$$CH_4 + H_2O \rightarrow 3H_2 + CO$$

Some of the high temperature fuel cells described in Chapter 6 can use this carbon monoxide as a fuel. However, fuel cells using platinum as a catalyst most certainly cannot! Even very small amounts of carbon monoxide have a very great affect on the anode. If a reformed hydrocarbon is to be used as a fuel, the carbon monoxide must be "shifted" to carbon dioxide using more steam:-

$$CO + H_2O \rightarrow H_2 + CO_2$$

This reaction is called the "water gas shift reaction". It does not easily go to completion, and there will nearly always be some carbon monoxide left in the reformed gas stream. A "state of the art" system will still have CO levels on the order of 0.25 to 0.5 % (= 2500 – 5000 ppm) (Cross, 1999).

The effect of the carbon monoxide is to occupy platinum catalyst sites – the compound has an affinity for platinum, and it covers the catalyst, preventing the hydrogen fuel from reaching it. Experience suggests that a concentration of carbon monoxide as low as 10 ppm has an unacceptable effect on the performance on a PEM fuel cell. This means that the CO levels in the fuel gas stream need to be brought down by a factor of 500 or more.

The methods used for removal of carbon monoxide from reformed fuel gas streams is discussed in some detail in Section 7.4.9, in Chapter 7. The extra processing needed adds considerably to the cost and size of a PEMFC system.

In some cases the requirement to remove carbon monoxide can be made somewhat less rigorous by the addition of small quantities of oxygen or air to the fuel stream (Stumper et al., 1998). This reacts with the carbon monoxide at the catalyst sites, thus removing it. Reported results show, for example, that adding 2% oxygen to a hydrogen gas stream containing 100 ppm carbon monoxide eliminates the poisoning effect of the carbon monoxide. However, any oxygen not reacting with CO will certainly react with hydrogen, and thus waste fuel. Also, the method can only be used for CO concentrations of 10s or 100s of ppm, not the 1000s of ppm concentration from typical fuel reformers. In addition, the system to feed the precisely controlled amounts of air or oxygen will be fairly complex, as the flow rate will need to carefully follow the rate of hydrogen use.

Typically, a gas clean-up unit will be used on its own, or a less effective (and lower cost) gas clean-up unit combined with some air feed to the fuel gas.

Another important point to note is that as the molecule length of the hydrocarbon fuel to be reformed becomes longer, the worse the problem becomes. With methane, CH_4, the initial reaction produced three hydrogen molecules for each CO molecule to be dealt with. If we go to a fuel such as C_8H_{18} then the initial reaction is:-

$$C_8H_{18} + 8H_2O \rightarrow 17H_2 + 8CO$$

The ratio of H_2 to CO is now about 2:1 - significantly more CO to be dealt with.

4.8.2   Methanol and other liquid fuels

An ideal fuel for any fuel cell would be a liquid fuel already in regular use, such as petrol. Unfortunately such fuels simply do not react at a sufficient rate to warrant consideration – with the exception of methanol. Methanol is already used as a fuel in some vehicles (e.g. certain types of racing car), and is readily available. It is used, for example, at the staggering rate of 250 million US gallons a year as a cleaner/solvent for automobile windscreens[3]. Methanol reacts at the anode of a PEM fuel cell, albeit slowly, according to the equation:-

$$CH_3OH + H_2O \rightarrow 6H^+ + 6e^- + CO_2$$

We note that the methanol needs to be mixed with water, and that six electrons are produced for each methanol molecule. Such fuel cells are called "direct methanol" fuel cells. This is in contrast to a fuel cell that uses hydrogen produced from methanol in a reformer such as that mentioned above in Section 2.8.1, which could be called an "indirect methanol" fuel cell. (The Gibbs free energy of this reaction was mentioned in Section 2.2.)

The most pressing problem with the direct methanol fuel cell is that the rate of reaction is very slow. This manifests itself as a high activation overvoltage on the anode. The result is a very low operating voltage, and thus a very low efficiency. The problem of fuel crossover (see Section 3.5) is also much worse than with hydrogen, as the methanol is absorbed into the polymer electrolyte.

Despite these problems, the easy availability, ease of storage and handling, high energy density and safety of methanol fuel make it so attractive that certain low power applications seem certain to appear soon. A good example might be a unit for recharging portable telephones. A direct methanol fuel cell could be incorporated into the holder of a cell phone, and keep it charged up while on standby. It has been calculated that such a unit would only have to have an efficiency of 4% to make it compete with battery technologies such as lithium.(Hockaday & Navas, 1999)

4.8.3   Using pure oxygen in place of air

It will be very rare for a designer to have a real choice between the use of air or oxygen in a PEM fuel cell system. Oxygen is used in air independent systems such as submarines and spacecraft, otherwise air is used. However, the use of oxygen does markedly improve the performance of a PEM fuel cell. This results from three effects:-

* The "no loss" open circuit voltage rises due to the increase in the oxygen partial pressure, as predicted by the Nernst equation, as noted in Section 2.5
* The activation overvoltage reduces, due to better use of catalyst sites, as noted in Section 3.4.3.
* The limiting current increases, thus reducing the mass transport or concentration overvoltage losses. This is because of the removal of the nitrogen gas, which is a major contributor to this type of loss at high current densities (see Section 3.7).

---

[3] Figure supplied by the American Methanol Institute in 1999

Published results (e.g. Prater, 1990) suggest that the effect of a change from air to oxygen will increase the power of a PEM fuel cell by about 30%. However, this will vary depending on the design of the cell. For example, a stack designed with poor reactant air flow will benefit more from a switch to oxygen.

Before we leave the topic of PEM fuel cells, two example systems will be presented. One is extremely simple, the other more complex. Both operate at approximately air pressure, and use pure hydrogen as the fuel, and air as the oxidant.

4.9   Example Systems

It will be apparent that there are many parameters of PEM fuel cells that can be changed. These changes, such as operating temperature and pressure, air stoichiometry, reactant humidity, water retaining properties of the gas diffusion membrane, and so on, affect the performance of the cell in fairly predictable ways. This naturally lends itself very well to the construction of computer models. This can be done for things as specific as the water and heat movement, alternatively the performance of a single cell under different conditions can be modelled (e.g. Thirumalai and White, 1997 and Wohn et al., 1998). Given the performance of a single cell, the performance of a whole stack can also be modelled (e.g. Lee and Lalk, 1998). An infinite variety of designs can thus be tried. However, by way of example, two PEM fuel cell systems are presented here to illustrate the issues we have been discussing in this chapter.

4.9.1   Small 12 Watt system

The first is a small 12 watt fuel cell designed for use in remote conditions, such as when camping, in boats, for military applications and for remote communications, and is shown in Figure 4.24. The electrodes are in the form of a disc or ring, as in Figure 4.25. The hydrogen feed is up a central tube. Strategically placed gaps in this tube feed the fuel to the anodes on the underside of each MEA. The top of each MEA has a thick gas diffusion layer which allows oxygen to diffuse in, and water to diffuse out to the edge, where it evaporates. The entire periphery is exposed to the air, so there is sufficient air circulation without any need for fans or blowers. Underneath the anode there is a gas diffusion layer, and an enclosure to prevent escape of hydrogen. This will typically be about 1.5 mm deep, and made of stainless steel.

The general construction method has features in common with the cell already described and shown in Figure 4.19, except that the gas diffusion layer is a more open material, but thicker, and extends up to the base of the next cell and thus also serves as the cell interconnect. The radial symmetry of the design simplifies manufacture. Such cells have been described by Daugherty et al. (1999) and Bossel (1999), and that shown in Figure 4.24 is made along these lines.

The hydrogen supply is "dead-ended", there is no circulation or venting – though there may be a system of periodic manual purging to release impurities from time to time.

Figure 4.24 Twelve watt fuel cell with no need for ventilation fans or other moving parts, as developed by DCH Technology. The fuel cell is the small cylinder in the middle of the picture. It is intended for uses such as camping, and is shown connected to a CD player, radio, and lamp. The connection to these devices is via a DC/DC converter (see Chapter 9).

The air circulation is basically self-regulating. When more power is drawn, more heat and more water are produced, both of which will decrease the density of the air, and thus increase circulation. On the other hand, because the air access to the cathode is limited, with water having to diffuse to the edge, the MEA does not dry out when the current is low. Although it does have the ability to respond to changes in power demand, the cell is well suited to continuous operation, and can very effectively be used in parallel with a rechargeable battery, which could provide higher power, though obviously only for short periods.

This then is an attempt to produce a "pure" fuel cell system, with the absolute minimum of extra devices. The efficiency, as with all fuel cells, varies with current, but at rated power it is liable to be around 48% (ref. LHV), i.e. the voltage of each cell will be about 0.6 volts.

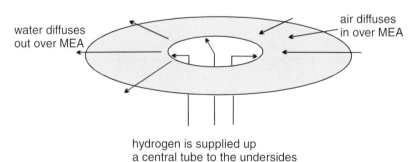

Figure 4.25 Principle of the 'disc type' PEM fuel cell

A way of indicating the various energy flows and power losses in a system is the 'Sankey diagram'. They are useful in all energy conversion systems, including fuel cells. This very "pure" fuel cell is a good place to introduce them, as it gives almost the simplest possible diagram, which is shown in Figure 4.26 below. Twenty five Watts of hydrogen power is the supply, which is converted into 12W of electrical power. The losses are simply 13 W of heat – there are no others. The losses are swept up or down out of the left to right line of the diagram.

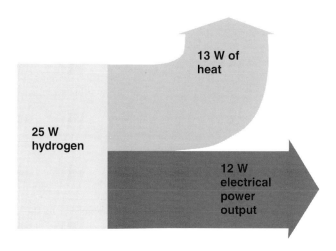

Figure 4.26 Sankey diagram for the very simple 'disc type' 12 W fuel cell.

4.9.2   Medium 2 kW system

The second fuel cell to be discussed here is a 2.0kW demonstrator designed and built by the Paul Scherrer Institute in Switzerland. A photograph is shown overleaf in Figure 4.29, a Sankey diagram in Figure 4.28, but we start with the system diagram in Figure 4.27 below. The fuel cell stack is made using the standard bipolar plate method. The bipolar plates are compression moulded carbon/polymer mixture (as in Section 4.6.2), with separate air cooling, as in Figure 4.16. The need to separate the cooling air and reactant air was explained in Section 4.5.2. To increase the maximum power of the fuel cell the reactant air is humidified, using the exit air as illustrated in Figure 4.14. The hydrogen gas is not simply "dead-ended", but rather is circulated using a pump. This is to help with the even humidification of the larger fuel cell stack. The circulation of the hydrogen is also a way of clearing impurities out of the cell.

In this case the Sankey diagram is rather more interesting. The final output power is just 1.64 kW. 1.8 kW of heat losses is shown, together with 360 W of electrical loss being swept downwards. This 360 W is needed to drive the three blowers and pumps, and to supply an electronic controller.

Although the reactant air flow will be less than the cooling air flow, it is against a pressure of about 0.1 bar, 10 kPa above atmospheric. This is due to length and narrowness of the reactant air path through the stack and humidifier. This is not a high pressure, but, as we shall see in Chapter 8, makes a big difference to the blower that can be used and the power consumed. The electric power consumed by the reactant air blower is 200 W, whereas for the cooling air only 70 W is needed. The hydrogen gas flow-rate is of course much less than that for the reactant air, and the pump only requires 50 W. All these pumps need controlling, and so the system has an electronic controller, which consumes 40 W. The electrical power lost amounts to 18% of the electrical power generated by the fuel cell. The net output power is 1.64 kW, making an efficiency of 42.5% (ref LHV).

The photograph of this system in Figure 4.29 allows these various parts of the system to be seen. It is a very clearly laid out system, which illustrates many points of good PEM fuel cell system design practice.

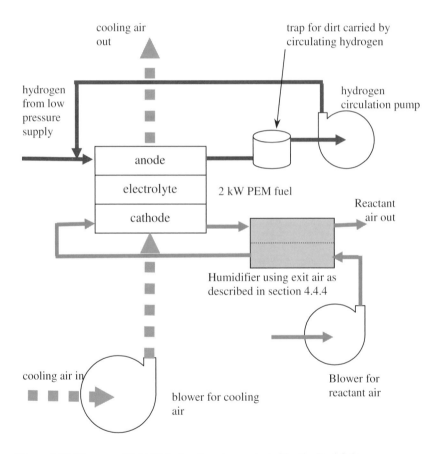

Figure 4.27 Diagram of 2.0 kW fuel cell as demonstrated by the Paul Scherrer Institute, Switzerland.

Figure 4.29 (On facing page) Photograph of 2 kW fuel cell system as demonstrated by the Paul Scherrer Institute. (Reproduced by kind permission of PSI.)

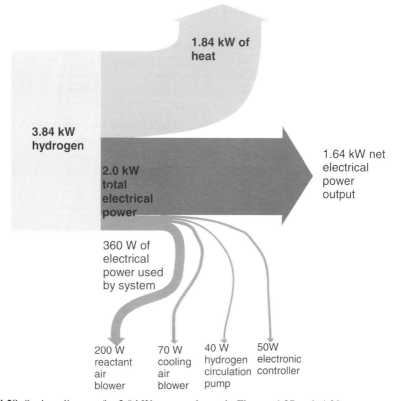

Figure 4.28 Sankey diagram for 2.0 kW system shown in Figures 4.27 and 4.29

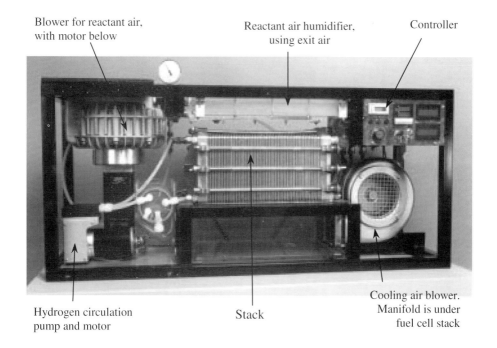

4.9.3   Large 260 kW system

Among the largest PEM fuel cell systems demonstrated is the fuel cell engine for buses produced by Ballard Power Systems, and shown in Figure 4.30. Buses are particularly suitable for fuel cell operation. Because they all refuel at one place – the bus depot – they can cope with novel fuels. Also, they operate in city centres, where pollution is often a major issue.

The fuel cell in this case has a power of about 260 kW, and the final motor drive is built into the system – hence the manufacturers call this a "fuel cell engine". The stacks are made using "mainstream" methods, that is using graphite bipolar plates. They are basically in two halves, as can be seen from Figure 4.30 below. Each half produces 130 kW, and consists of about 750 cells in series. They are water cooled. At maximum power the system is pressurised to 3 bar. The compression system and other ancillaries consume about 60 kW of power, and are mounted between the two main stacks.

Figure 4.30 Fuel cell engine that fits in the place normally taken by the diesel engine. The power of the fuel cell is about 260 kW. (Photograph by kind permission of Ballard Power Systems.)

This fairly complex high power system uses compressors of the type discussed in Chapter 8, and various electrical sub-systems discussed in Chapter 9. So we will leave this system as a 'stepping stone' to those later chapters, and we will look at it again in more detail in Section 9.6 near the end of the book. Readers who are only interested in PEMFCs could now move on to Chapter 7, where the very important question of "where does the hydrogen come from?" is addressed. In Chapters 5 and 6 we consider other fuel cell types in more detail.

References

Barbir, F., (1999) "Fuel Cell Powered Utility Vehicles", Proceedings of the European Fuel Cell Forum Portable Fuel Cells conference, Lucerne, pp113-121

Bevers D., Wagner N., VonBradke M. (1998) "Innovative production procedure for low cost PEFC electrodes and electrode/membrane structures", *International Journal of Hydrogen Energy,* Vol.23, No.1, pp57-63

Bossel U.G. (1999) "Portable Fuel Cell Charger with Integrated hydrogen generator", Proceedings of the European Fuel Cell Forum Portable Fuel Cells conference, Lucerne, pp79-84

Büchi F.N., Srinivasan S., (1997) "Operating proton exchange membrane fuel cells without external humidification of the reactant gases. Fundamental aspects" *Journal of the Electrochemical Society,* Vol 144, No 8, pp2767-2772

Cross J.C., (1999) "Hydrocarbon reforming for fuel cell application", Proceedings of the European Fuel Cell Forum Portable Fuel Cells Conference, Lucerne, pp205-213

Daly B.B. (1979) *Woods Practical Guide to Fan Engineering,* Woods of Colchester Ltd.

Daugherty M., Haberman D., Stetson N., Ibrahim S., Lokken D., Dunn D., Cherniak M., Salter C. (1999) "Modular PEM fuel cell for outdoor applications", Proceedings of the European Fuel Cell Forum Portable Fuel Cells conference, Lucerne, pp69-78

Giorgi L., Antolini E., Pozio A., Passalacqua E. (1998) "Influence of the PTFE content in the diffusion layer of low-Pt loading electrodes for polymer electrolyte fuel cells", *Electrochimica Acta,* Vol 43, No.24, pp3675-3680

Heinzel A. et al.(1998) "Membrane fuel cells – concepts and system design", *Electrochimica Acta,* Vol.43, No.24, pp3817-3820 Flat PEM fuel cells

Hirschenhofer J.H., Stauffer D.B., & Engleman R.R.(1995) *"Fuel Cells: A Handbook",* Revision 3, Business/Technology Books

Hockaday R. & Navas C. (1999) "Micro-Fuel Cells for Portable Electronics", Proceedings of the European Fuel Cell Forum Portable Fuel Cells Conference, Lucerne, pp45-54

Kiefer J., Brack H-P., Huslage J., Buchi F.N., Tsakada A., Geiger F., Schere G.G. (1999) "Radiation grafting: a versatile membrane preparation tool for fuel cell applications", Proceedings of the European Fuel Cell Forum Portable Fuel Cells conference, Lucerne, pp227-235

Koppel T. (1999) *Powering the future - the Ballard fuel cell and the race to change the world,* John Wiley & Sons.

Lee J.H., Lalk T.R. (1998) "Modelling fuel cell stack systems", *Journal of Power Sources,* Vol. 73, No.2, pp229-241

Lee S.J. (1998) "Effects of nafion impregnation on performance of PEMFC electrodes", *Electrochimica Acta,* Vol 43, No.24, pp3693-3701 (rolling of electrodes, preparation of MEAs)

Murphy O.J., Cisar A., Clarke E. (1998) "Low cost light weight high power density PEM fuel cell stack", *Electrochimica Acta,* Vol 43, No.24, pp3829-3840

Okada T., Xie G., Meeg M., (1998) "Simulation for water management in membranes for polymer electrolyte fuel cells", *Electrochimica Acta,* Vol.43,No 14-15, pp2141-2155

Prater K.,(1990) "The renaissance of the solid polymer fuel cell", *Journal of Power Sources,* Vol.29, Nos1&2, pp239-250

Ralph T.R., Hards G.A., Keating J.E. Campbell S.A., Wilkinson D.P., Davis H., StPierre J., Johnson M.C. (1997) "Low cost electrodes for proton exchange membrane fuel cells", *Journal of the Electrochemical Society,* Vol.144, No 11, pp3845-3857

Stumper J., Campbell S.A., Wilkinson D.P., Johnson M.C., Davis M. "In-situ methods for the determination of current distributions in PEM fuel cells", *Electrochimica Acta,* Vol.43, No.24, pp3773 - 3783

Thirumalai D. & White R.E. (1997) "Mathematical modelling of proton-exchange-membrane fuel-cell stacks", *Journal of the Electrochemical Society*, Vol. 144, No.5, pp1717-1723

Warshay M., Prokopius P.R., Le M., Voecks G. (1997) "NASA fuel cell upgrade program for the Space Shuttle Orbiter", *Proceedings of the Intersociety Energy Conversion Engineering Conference*, Vol.1, pp228-231

Warshay M., & Prokopius P.R., (1990) "The fuel cell in space: yesterday, today and tomorrow", *Journal of Power Sources*, Vol.29, Nos 1&2, pp193-200

Watanabe M., Uchida H., Seki Y., Emon M., Stonehart P. (1996) "Self humidifying polymer electrolyte membranes for fuel cells", *Journal of the Electrochemical Society*, Vol 143, No 12, pp3847-3852,

Wilson M.S., Gottesfield S. (1992) "High performance catalysed membranes of ultr-low Pt loadings for polymer electrolyte fuel cells" *Journal of the Electrochemical Society*, Vol 139, No 2, L28-30,

Wohn M., Bolwin K., Schnurnberger W., Fisher M., Neubrand W., Eigernberger G. (1998) "Dynamic modelling and simulation of a polymer membrane fuel cell including mass transport limitations", *International Journal of Hydrogen Energy*, Vol.23, No.3, pp213-218,

Wood D.L., Yi J.S., Nguyen T.V. (1998) "Effect of direct liquid water injection and interdigitated flow field on the performance of proton exchange membrane fuel cells", *Electrochimica Acta,* Vol.43, No.24, pp3795-3809

Yeager H.J., Eisenberg A. (Eds.) "Perflourinated ionomer membranes", ACS Symp. Ser.No. 189, American Chemical Society

Zawodzinski T.A., Derouin C., Radzinski S., Sherman R.J., Smith V.T., Springer T.E., Gottesfield S. (1993) "Water uptake by and transport through Nafion 117 membranes", *Journal of the Electrochemical Society*, Vol.140, no.4, pp1041-1047

5

Alkaline Electrolyte Fuel Cells

5.1   Historical Background and Overview

The basic chemistry of the alkaline electrolyte fuel cell (AFC) was explained in Chapter 1, Figure 1.4. The reaction at the anode is:-

$$2H_2 + 4OH^- \rightarrow 4H_2O + 4e^-$$ [5.1]

The electrons released in this reaction pass round the external circuit, reaching the cathode, where they react, forming new OH^- ions.

$$O_2 + 4e^- + 2H_2O \rightarrow 4OH^-$$ [5.2]

The alkaline fuel cell has been described since at least 1902[1], but they were not demonstrated as viable power units till the 1940s and 1950s by F.T.Bacon at Cambridge, England. Although the PEM fuel cell was chosen for the first NASA manned spacecraft, it was the alkaline fuel cell that took man to the moon with the Apollo missions. The success of the alkaline fuel cell in this application, and the demonstration of high power working fuel cells by Bacon, led to a good deal of experiment and development of alkaline fuel cells during the 1960s and early 1970s. Demonstration alkaline fuel cells were used to drive agricultural tractors (Ihrig, 1960), power cars (Kordesh, 1971), provide power to offshore navigation equipment and boats, drive fork lift trucks, and so on. Descriptions and photographs of these systems can be found in McDougall (1976) and Kordesh & Simader (1996), as well as the papers already referred to. Although many of these systems worked reasonably well as demonstrations, other difficulties, such as cost, reliability, ease of use, ruggedness and safety were not so easily solved. When attempts were made to solve these wider engineering problems, it was found that, at that time, fuel cells could not compete with rival energy conversion technologies, and research and development was scaled down. The success of PEM fuel cell developments in recent years has furthered the decline in interest in alkaline fuel cells, and now very few companies or research groups are working in the field. The space program remained the one shining star in the alkaline fuel cell world, with the Apollo system being improved and developed for the space shuttle Orbiter vehicles. However, the fact that the new fuel cells for the space shuttle

---

[1] J.H.Reid, US patent no. 736 016 017 (1902)

Figure 5.1 1.5 kW fuel cell from the Apollo spacecraft. Two of these units were used, each weighing 113 kg. (Photograph by kind permission of International Fuel Cells.)

Orbiter vehicles will be PEMFCs (Warshay, 1997) only serves to further underline their decline in importance. Nevertheless, because of their success with the space program, alkaline fuel cells played a hugely important role in keeping fuel cell technology development going through the later half of the 20th century. Also, it could well be that the problems that we will be discussing in this chapter can be solved. For example, if the major source of hydrogen is electrolysed water from the use of solar panels, then alkaline fuel cells could well become more attractive.

The major advantages of alkaline electrolyte fuel cells are that the activation overvoltage at the cathode (discussed in Chapter 3) is generally less than with an acid electrolyte. Also, the electrodes can be made from non-precious metal electrodes, and no particularly exotic materials are needed.

The electrolyte obviously needs to be an alkaline solution. Sodium hydroxide and potassium hydroxide solution, being the lowest cost, highly soluble, and not excessively corrosive, are the prime candidates. It turns out that potassium carbonate is much more soluble than sodium carbonate, and as we shall see in Section 5.5, this is an important advantage. Since the cost difference is small, potassium hydroxide solution is *always* used as the electrolyte. However, as we shall see, this is about the only

Figure 5.2 12 kW fuel cell used on the Space Shuttle Orbiter. This unit weighs 120 kg, and measures approximately 36×38×114 cm. The nearer part is the control system, pumps etc, and the stack of 32 cells in series is under the white cover. Each cell operates at about 0.875 volts. (Photograph by kind permission of International Fuel Cells.)

common factor among different AFCs. Other variables such as pressure, temperature and electrode structure, vary greatly between designs. For example, the Apollo fuel cell operated at 260 °C, whereas the temperature of the Orbiter alkaline fuel cell is about 90 °C. The different arrangements for dealing with the electrolyte, the different types of electrodes used, different catalysts, the choice of pressure and temperature and so on are discussed in the sections that follow.

5.2   Types of Alkaline Electrolyte Fuel Cell

5.2.1   Mobile electrolyte

The basic structure of the mobile electrolyte fuel cell is shown below in Figure 5.3. The KOH solution is pumped around the fuel cell. Hydrogen is supplied to the anode, but must be circulated, as it is at the anode that the water is produced. The hydrogen will evaporate the water, which is then condensed out at the cooling unit that the hydrogen is circulated through. The hydrogen comes from a compressed gas cylinder, and the circulation is achieved using an ejector circulator, as described in Chapter 8, Section 8.10. The system shown in Figure 5.3 uses air, rather than oxygen. As well as the desired fuel cell reactions of equations 5.1 and 5.2, the carbon dioxide in the air will react with the potassium hydroxide electrolyte:-

$$2\,KOH \;+\; CO_2 \;\rightarrow\; K_2CO_3 + H_2O \qquad\qquad\qquad [5.3]$$

The potassium hydroxide is thus gradually changed to potassium carbonate. The effect of this is that the concentration of OH^- ions reduces as they are replaced with carbonate CO_3^{2-} ions, which greatly affects the performance of the cell. This major difficulty is discussed further in Section 5.5, but one way of at least reducing it is to remove as much as possible of the CO_2 from the air, and so this is done in the cathode air supply system using a CO_2 scrubber.

The disadvantages of the mobile electrolyte centre around the extra equipment needed. A pump is needed, and the fluid to be pumped is corrosive. The extra pipework means more possibilities for leaks, and the surface tension of KOH solution makes for a fluid that is prone to find its way through the smallest of gaps. Also, it becomes harder to design a system that will work in any orientation. There is a further very important problem cannot be deduced from Figure 5.3, since it only shows one cell. The circulation system should give the longest and narrowest possible current path, and preferably isolation, between the KOH solution in each cell - otherwise the electrolyte of each and every cell will be connected together, and there will be an internal "short-circuit"[2].

However, there are many advantages to the mobile electrolyte system. The principle benefits are:-

- The circulating electrolyte can serve as a cooling system for the fuel cell.

---

[2] This internal current can be measured by finding the hydrogen consumption at open circuit. For example, in the cells used by Kordesh (1971), the internal current was found to be about 1.5 mA.cm^{-2}.

- The electrolyte is continuously stirred and mixed. It will be seen from the anode and cathode equations given at the beginning of the chapter that water is consumed at the cathode, and produced (twice as fast) at the anode. This can result in the electrolyte becoming too concentrated at the cathode – so concentrated that it solidifies. Stirring reduces this problem.
- Having the electrolyte circulate means that if the product water transfers to the electrolyte, rather than evaporating at the anode, then the electrolyte can be passed through a system for restoring the concentration (i.e. an evaporator).
- It is comparatively straightforward to pump out all the electrolyte, and replace it with a fresh solution, if it has become too diluted by reaction with carbon dioxide, as in equation 5.3.

This mobile electrolyte system was used by Bacon in his historic alkali fuel cells of the 1950s, and in the Apollo mission fuel cells, among others. However, the Orbiter vehicles used a static electrolyte, as described in the next section.

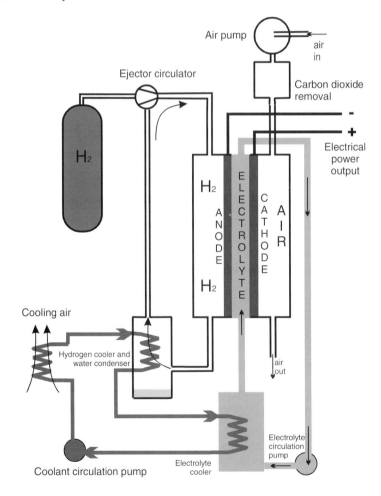

Figure 5.3 Diagram of an alkaline fuel cell with mobile electrolyte. The electrolyte also serves as the fuel cell coolant.

5.2.2   Static electrolyte alkaline fuel cells

An alternative to a "free" electrolyte, which circulates as in Figure 5.3 is for each cell in the stack to have their own, separate, electrolyte which is held in a matrix material between the electrodes. Such a system is shown in Figure 5.4. The KOH solution is held in a matrix material, which is usually asbestos. This material has excellent porosity, strength, and corrosion resistance, although of course its safety problems would be a difficulty for a fuel cell system designed for use by members of the public.

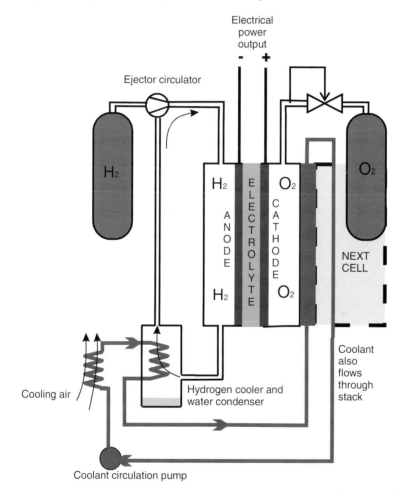

Figure 5.4 Alkaline electrolyte fuel cell with static electrolyte held in a matrix. This system uses pure oxygen instead of air.

The system of Figure 5.4 uses pure oxygen at the cathode, though that is not obligatory for a matrix held electrolyte. The hydrogen is circulated as with the previous system, in order to remove the product water. In the spacecraft systems this product water is used for drinking, cooking and cabin humidification. However, a cooling system will also be needed, and so cooling water, or other fluid, is needed. In the Apollo system it was a glycol/water mixture, as is used in car engines. In the Orbiter systems the cooling fluid is a fluorinated hydrocarbon dielectric liquid (Warshay & Prokopius, 1990).

This matrix held electrolyte system is essentially like the PEM fuel cell – the electrolyte is to all intents and purposes solid, and can be in any orientation. A major advantage is of course that the electrolyte does not need to be pumped around or "dealt with" in any way. There is also no problem of the internal "short circuit" which can be the result of a pumped electrolyte. However, there is the problem of managing the product water, the evaporation of water, and the fact that water is used at the cathode. Essentially the water problem is similar to that for PEM fuel cells, though "inverted", in that water is produced at the anode, and removed from the cathode. (In the PEMFC water is produced at the cathode, and removed from the anode by electro-osmostic drag, as explained in Section 4.4.) The fuel cell must be designed so that the water content of the cathode region is kept sufficiently high by diffusion from the anode. Generally though, the problem of water management is much less severe than with the PEMFC. For one thing, the saturated vapour pressure of KOH solution does not rise so quickly with temperature as it does with pure water, as we will see in Section 5.4 below. This means that the rate of evaporation is much less.

In space applications the advantages of greater mechanical simplicity mean that this approach is now used. However, for terrestrial applications, where the problem of carbon dioxide contamination of the electrolyte is bound to occur, renewal of the electrolyte must be possible. For this matrix type cell this would require a complete fuel cell rebuild operation. Also, the use of asbestos is a severe problem, as it is hazardous for health, and in some countries its use is banned. A new material needs to be found, but research in this area in unlikely to be undertaken while the likelihood of its eventual use is so low.

5.2.3   *Dissolved fuel alkaline fuel cells*

This type of fuel cell is unlikely to be used for serious power generation, but is included as it is the simplest type of fuel cell to manufacture, and it shows how the alkaline electrolyte can be the basis of really very simple fuel cells. It has been used in a number of successful fuel cell demonstrators.

The principle is shown in Figure 5.5. The electrolyte is KOH solution, with a fuel, such as hydrazine, or ammonia, mixed with it. The fuel anode is along the lines discussed in Section 5.3.4 below, with a platinum catalyst. The fuel is also fully in contact with the cathode. This makes the "fuel crossover" problem discussed in Section 3.5 very severe. However, in this case it does not matter greatly, as the cathode catalyst is not platinum, and so the rate of reaction of the fuel on the cathode is very low. There is thus only one seal that could leak, a very low pressure joint around the cathode. The cell is re-fuelled simply by adding more fuel to the electrolyte.

An ideal fuel for this type of fuel cell is hydrazine, H_2NNH_2, as it dissociates into hydrogen and nitrogen on a fuel cell electrode. The hydrogen then reacts normally. (Other reaction routes with similar results are possible, as described by McDougall (1976), p.71.) Low cost, compact, simple and easily re-fuelled cells can be made this way. Unfortunately however, hydrazine is toxic, carcinogenic, and explosive! It is used as a fuel in certain very well regulated circumstances, but is not suitable for widespread use, and so cells using it can only be for demonstration purposes.

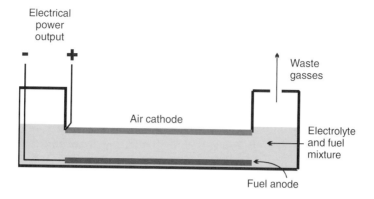

Figure 5.5 Dissolved fuel fuel cell, arguably the simplest of all types.

This dissolved fuel type of cell can be used with other liquid fuels, such as methanol. In this case the anode reaction is:-

$$CH_3OH + 6OH^- \rightarrow 5H_2O + CO_2 + 6e^-$$

So, as well as electrons, carbon dioxide is produced. This reacts with the KOH solution, as shown above in equation 5.3, converting it to carbonate. This effect makes the cell impractical for use a power source. Indeed, since the electrolyte is "used up" by the cell reaction, it could be argued that the system is not a true fuel cell.

The dissolved fuel principle can in theory, be used with acid electrolyte fuel cells. However, practical considerations mean that only alkali electrodes are viable. It is very difficult, for example, to make an active catalyst in a low temperature acid electrolyte fuel cell that does not use precious metals, and that will not therefore oxidise the fuel.

Figure 5.6 This simple, very low cost cell is of the "dissolved fuel" type. It is designed for use in education and for demonstration, and works with methanol, hydrazine and other liquid fuels

5.3   Electrodes for Alkaline Electrolyte Fuel Cells

5.3.1   Introduction

We have already pointed out that alkaline fuel cells can be operated at a wide range of temperatures and pressures. It is also the case that their range of applications is quite restricted. The result of this is that there is no standard type of electrode for the AFC, and different approaches are taken depending on performance requirements, cost limits, operating temperature and pressure. Different catalysts can be used also, but this does not necessarily affect the electrode structure. For example, platinum catalyst can be used with any of the main electrode structures described here.

5.3.2   Sintered nickel powder

When F.T.Bacon designed his historically very important fuel cells in the 1940s and 1950s he wanted to use simple, low cost materials, and avoided precious metal catalysts. He thus opted for nickel electrodes. These were made porous by fabricating them from powdered nickel, which was then sintered to make a rigid structure. To enable a good three phase contact between the reactant gas, the liquid electrolyte and the solid electrode, the nickel electrode was made in two layers using two sizes of nickel powders. This gave a wetted fine pore structure for the liquid side, and more open pores for the gas side. (Appleby, 1990) This structure gave very good results, though careful control of the differential pressure between the gas and the electrolyte was needed to ensure that the liquid gas boundary sat at the right point, as wet proofing materials like PTFE were not available at that time.

   This electrode structure was also used in the Apollo mission fuel cells. Such structures may or may not be combined with catalysts. In the Apollo and Bacon cells the anode was formed from the straightforward nickel powder as described above, whereas the cathode was partially lithiated and oxidised.[3]

5.3.3   Raney metals

An alternative method of achieving a very active and porous form of a metal, which has been used for alkaline fuel cells from the 1960s through to the present, is the use of Raney metals. These are prepared by mixing the active metal (for example nickel) with an inactive metal, usually aluminium. The mixing is done in such a way that distinct regions of aluminium and the host metal are maintained - it is not a true alloy. The mixture is then treated with a strong alkali which dissolves out the aluminium. This leaves a porous material, with very high surface area. This process does not require the sintering of the nickel powder method, yet does give scope for changing the pore size, since this can be varied by altering the degree of mixing of the two metals.

   Raney nickel electrodes prepared in this way were used in many of the fuel cell demonstrations mentioned in the introduction to this chapter. Often Raney nickel was used for the anode, and silver for the cathode. This combination was also used for the electrodes

---

[3] The addition of lithium salts to the surface of the cathode reduced oxidation of the nickel

of the Siemens alkaline fuel cell used in submarines in the early 1990s (Strasser, 1990). They have also been used more recently, in a ground up form, in the rolled electrodes to be described in Section 5.3.4 below. (Gulzow 1996).

5.3.4   Rolled electrodes

Modern electrodes tend to use carbon supported catalysts, mixed with PTFE, which are then rolled out onto a material such as nickel mesh. The PTFE acts as a binder, and also its hydrophobic properties stop the electrode flooding, and provide for controlled permeation of the electrode by the liquid electrolyte. A thin layer of PTFE will often be put over the surface of the electrode to further control the porosity, and to prevent the electrolyte passing through the electrode, without the need to pressurise the reactant gases, as has to be done with the porous metal electrodes. Carbon fibre is sometimes added to the mix to increase strength, conductivity and roughness.

The manufacturing process can be done using modified paper making machines, at quite low cost. Such electrodes are not just used in fuel cells, but are also used in metal/air batteries, for which the cathode reaction is much the same as for an alkali fuel cell. For example, the same electrode can be used as the cathode in a zinc air battery (e.g. for hearing aids), an aluminium/air battery (e.g. for telecommunications reserve power), and an alkaline electrolyte fuel cell. Such an electrode is shown in Figure 5.7 below. The carbon supported catalyst is of the same structure as that shown in Figure 4.6 in the previous chapter. However, the catalyst will not always be platinum. For example, manganese can be used for the cathode in metal air batteries and fuel cells.

Figure 5.7 Photograph showing the structure of a rolled electrode. The catalyst is mixed with a PTFE binder and rolled onto nickel mesh. The thin layer of PTFE on the gas side is shown partially rolled back.

With a non-platinum catalyst such electrodes are readily available at a cost from about $0.01 per cm^2, or around $10 per ft^2, i.e. very low cost compared to other fuel cell materials. Adding a platinum catalyst increases the cost, depending on the loading, but it might only be by a factor of about 3, which still gives, in fuel cell terms, a very low cost electrode. However, there are problems.

One problem is that the electrode is covered with a layer of PTFE, and so the surface is non-conductive, and thus a bipolar plate cannot be used for cell interconnection. Instead, the cells have to be edge connected. This is not so bad, as with the nickel mesh running right through the electrode, its surface conductivity is quite good. A more serious problem is the effect of carbon dioxide on the electrode performance. This is in addition to the effect that carbon dioxide has on the electrolyte, which we noted above in equation 5.3, though it might be connected, in that there could be an effect of contamination by the carbonate. In reviewing the literature in this area, in Hircschenhofer et al. (1995) it is noted that the lifetime of such carbon supported catalyst electrodes was from 1600 to 3400 hours at 65 °C and 100 $mA.cm^{-2}$ when using air containing CO_2. However, if the carbon dioxide is removed from the air the lifetime of the electrodes increases to at least 4000 hours under similar conditions. This is not a particularly high current density, and lifetime is less at higher currents. Lower temperatures also shorten life, presumably because the solubility of the carbonate decreases. 1600 hours is only about 66 days, and is not suitable for any but the smallest number of applications. (Note, however, that it is quite adequate for the operating time of a battery, for which this type of electrode is generally used.)

When using such carbon supported electrodes it is clear that the carbon dioxide must be removed from the air. Another approach to the problem is the use of a different type of rolled electrode, which does not use carbon supported catalyst. Gulzow (1996) describes an anode based on granules of Raney nickel mixed with PTFE. This is rolled onto a metal net in much the same way as the PTFE/carbon supported catalyst. A cathode can be prepared in much the same way, only using silver instead of nickel. It is claimed that these electrodes are not damaged by CO_2.

5.4   Operating Pressure and Temperature

Historically, the major alkaline electrolyte fuel cells have operated at well above ambient pressure and temperature. The pressure and temperature, together with information about the electrode catalyst, is given for a selection of important alkaline fuel cells in Table 5.1 below.

Table 5.1 Operating parameters for certain alkaline electrolyte fuel cells.[Data from Warshay & Prokopius (1990) and Strasser (1990)] The pressure figures are approximate, since there are usually small differences between each reactant gas.

Fuel cell	Pressure (bar)	Temp. (°C)	KOH (% conc.)	Anode catalyst	Cathode catalyst
Bacon	45	200	30	Ni	NiO
Apollo	3.4	230	75	Ni	NiO
Orbiter	4.1	93	35	Pt/Pd	Au/Pt
Siemens	2.2	80		Ni	Ag

The advantages of higher pressure have been seen in Chapter 2, where it was shown, in Section 2.5.4, that the open circuit voltage of a fuel cell is raised when the pressure increases according to the formula:-

$$\Delta V = \frac{RT}{4F} \ln \left(\frac{P_2}{P_1} \right)$$

The demonstration cell of F.T.Bacon operated at around 45 bar, and 200 °C. However, even these high pressures would only give a "lift" to the voltage of about 0.04 volts if this "Nernstian" effect was the sole benefit. We also saw, in Chapter 3, Section 3.4, that a rise in pressure (and/or temperature) increases the exchange current density, which reduces the activation overvoltage on the cathode. The result is that the benefit of increased pressure is much more than the equation above would predict. As a consequence, the very high pressure gave the "Bacon cell" a performance that even today would be considered remarkable – 400 mA.cm^{-2} at 0.85 volts, or 1 A.cm^{-2} at 0.8 volts.

The choice of operating pressure and temperature, the KOH concentration, and the catalyst used are all linked together. A good example is the movement from the Bacon cell to the system used in the Apollo spacecraft. As we have seen, the performance of the "Bacon cell" is very good, but it was a heavy engineering design, a very high pressure system. To obtain a low enough mass for space applications the pressure had to be reduced. However, to maintain the performance at an acceptable level, the temperature had to be increased. It was then necessary to increase the concentration of the KOH to 75%, otherwise the electrolyte would have boiled. Increasing the KOH concentration considerably lowers the vapour pressure, as can be seen from Figure 5.8. At ambient temperature this concentration of KOH is solid, and so the system had to be supplied with heaters to start the system. In the Orbiter system the electrolyte concentration was reduced back to 32%, and the temperature to 93 °C. To maintain performance an electroactive catalyst is needed, and 20 mg.cm^{-2} of gold/platinum alloy is used on the cathode, and 10 mg.cm^{-2} Pt on the anode.

Most alkaline electrolyte fuel cells operate using reactant gases from high pressure or cryogenic storage systems. In each case the gas is supplied at fairly high pressure, and so the "costs" involved with operating at high pressure are those connected with the extra mass involved with designing a high pressure system. As well as the problem of containing the gases, and preventing leaks, there is also the problem of internal stresses if the reactants are at different pressures, so the two pressures must be accurately controlled.

The problem of leaks from high pressure systems is obviously a concern. Apart from the waste of gas, there is also the possibility of the build up of explosive mixtures of hydrogen and oxygen, especially when the cell is for use in confined spaces such as submarines. One solution to this problem is to provide an outer envelope for the fuel cell stack that is filled with nitrogen, and to have this nitrogen at a higher pressure than any of the reactant gases. As an example, Strasser (1990) describes the Siemens system, where the hydrogen is supplied at 2.3 bar, the oxygen at 2.1 bar, and the surrounding nitrogen gas is at 2.7 bar. Any leak would result in flow of nitrogen into the cells, which would reduce the performance, but would prevent an outflow of reactant gas

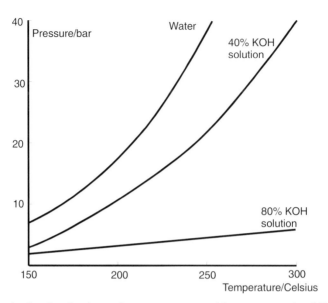

Figure 5.8 Graphs showing the change in vapour pressure with temperature for different concentrations of KOH solution.

In AFCs there is often a difference in the pressure of the reactant gases, and/or of the electrolyte. The pressure differences in the Siemens AFC, where the hydrogen pressure is slightly higher, have already been noted. In the Orbiter fuel cell the hydrogen gas pressure is kept at 0.35 bar *below* the oxygen pressure. In the Apollo system the gases were at the same pressure, but both were about 0.7 bar above the pressure of the electrolyte. No general rules can be given about this. The small pressure differences will be required for a variety of reasons – electrode diffusion and product water management being two examples. The needs will strongly depend on the details of the system.

We saw in Chapter 2 that raising the temperature actually reduces the "no loss" open circuit voltage of a fuel cell. However, in practice, this change is far exceeded by the reduction in the activation overvoltage, especially on the cathode. As a result, increasing the temperature increases the voltage from an AFC. Hirschenhofer et al.(1995) has shown, from a wide review of results, that below about 60 °C there is a very large benefit to raising the temperature, as much as 4 mV/ °C for each cell. At this rate, increasing the temperature from 30 to 60 °C would increase the cell voltage by about 0.12 volts, a big increase in the context of fuel cells operating at about 0.6 volts per cell. At higher temperatures there is still a noticeable benefit, but only in the region of 0.5 mV/ °C. We could conclude that about 60 °C would be a minimum operating temperature for an AFC. Above this, the choice would depend strongly on the power of the cell (and thus any heat losses) the pressure, and the effect of the concentration of the electrolyte on the rate of evaporation of water.

5.5   Problems and Development

We have already noted that activity in the area of alkaline fuel cells is currently very low. The main problem with AFCs for terrestrial applications is the problem of carbon dioxide reactions with the alkaline electrolyte. This occurs with the carbon dioxide in the air, and would happen even more strongly if hydrogen derived from hydrocarbons (such as methane) were used as the fuel.

The problem is that the carbon dioxide reacts with the hydroxide ion, forming carbonate, as in equation 5.3. The effects of this are:-

- The OH^- concentration is reduced, thus reducing the rate of reaction at the anode.
- The viscosity is increased, reducing the diffusion rates, thus lowering the limiting currents and increasing the mass transport losses, as discussed in Chapter 3, Section 3.7
- The carbonate salt is less soluble, and so will eventually precipitate out, blocking the pores of the electrodes. (Note, this problem is much less severe with potassium carbonate, which is why KOH is used in preference to NaOH.)
- Oxygen solubility is reduced, increasing the activation losses at the cathode
- The electrolyte conductivity is reduced, increasing the ohmic losses
- The electrodes performance may be degraded, as noted above in Section 5.3.4

Most of the notable achievements of AFCs have been using pure hydrogen and oxygen supplies for air independent power sources. However, the scope for these is limited.

For an alkaline electrolyte fuel cell to work well over a long period, it is essential to remove the carbon dioxide from the air. This can be done, but of course increases costs, complexity, mass and size. One way that shows particular promise has been described by Ahuja and Green (1998), though it can only be used when hydrogen is stored cryogenically. Basically, it takes advantage of the fact that heat exchangers are needed to warm the hydrogen, and cool the fuel cell. These could be combined with a system to freeze the carbon dioxide out of the incoming air, which can then be reheated.

Another possibility, which is actually what Bacon had in mind when developing his AFC designs in the mid 20th century, is to incorporate the cells into a regenerative system. Electricity from renewable sources is used to electrolyse water when the energy is available, and the fuel cell turns it back into electricity as needed. Of course, any fuel cell could be used in such a system, but here the AFC's disadvantages would be largely removed, since both hydrogen and oxygen would be generated. Thus their advantages of low cost, simplicity, lack of exotic materials, good cathode performance, and wide range of operating temperatures and pressures, might bring them to the fore again. In the mean time, a few companies and research institutions are continuing to develop these devices.

References

Ahuja V. & Green R. (1998) "Carbon dioxide removal from air for alkaline fuel cells operating with liquid nitrogen – a synergistic advantage", *International Journal of Hydrogen Energy,* Vol.23, no 2, pp131-137

Appleby A.J. (1990) "From Sir William Grove to today, fuel cells and the future" *Journal of Power Sources*, vol.29, pp3-11

Gulzow E. (1996) "Alkaline fuel cells: a critical view" *Journal of Power Sources*, vol.61, pp99-104

Hirschenhofer J.H., Stauffer D.B. & Engleman R.R. (1995) *"Fuel Cells: a Handbook"*, revision 3, Business/Technology Books, pp6-10 to 6-15

Ihrig H.K.(1960), 11th Annual Earthmoving Industry Conference, SAE Paper No. S-253

Kordesh K.V. (1971) "Hydrogen-air/lead battery hybrid system for vehicle propulsion", *Journal of the Electrochemical Society,* Vol.118, no. 5, pp812 – 817

Kordesh K.V. & Simader G. (1996) *Fuel Cells and their Applications*, VCH Verlagsgesellschaft.

McDougall A. (1976) *Fuel Cells*, Macmillan

Strasser K. (1990) "The design of alkaline fuel cells" *Journal of Power Sources,* vol. 29, pp149-166

Warshay M. & Prokopius P.R. (1990) "The fuel cell in space: Yesterday, today and tomorrow" *Journal of Power Sources*, vol 29, pp193-200

Warshay M., Prokopius P.R., Le M., Voecks G. (1996) "NASA fuel cell upgrade program for the space shuttle orbiter", *Proceedings of the Intersociety Energy Conversion Engineering Conference,* Vol.1, pp1717-1723

6

Medium and High Temperature Fuel Cells

6.1   Introduction

In Chapter 2 we noted that the "no loss" open circuit voltage for a hydrogen fuel cell reduces at higher temperatures. Indeed, we noted that above about 800 °C the theoretical maximum efficiency of a fuel cell was actually less than that of a heat engine. It might therefore seem unwise to operate fuel cells at higher temperatures! However, these apparent problems are in many cases outweighed by the advantages of higher temperature. The main advantages are:-

- The electrochemical reactions proceed more quickly, which manifests in lower activation voltage losses. Also, noble metal catalysts are often not needed.

- The high temperature of the cell and the exit gases means that there is heat available from the cell at temperatures high enough to facilitate the extraction of hydrogen from other more readily available fuels, such as natural gas.

- The high temperature exit gases and cooling fluids are a valuable source of heat for buildings, processes and facilities near the fuel cell. In other words, such fuel cells make excellent "combined heat and power (CHP) systems".

- The high temperature exit gases and cooling fluids can be used to drive turbines, which can drive generators, producing further electricity. This is known as a "bottoming cycle". We will show in Section 6.2.4 below, that this combination of a fuel cell and a heat engine allows the complementary characteristics of each to be used to great advantage, providing very efficient electricity generation.

There are three types of medium and high temperature fuel cell that we shall be considering in this chapter. The phosphoric acid electrolyte fuel cell (PAFC) is the most well developed of the three. Many PAFC 200 kW combined heat and power systems are installed at hospitals, military bases, leisure centres, offices, factories, and even prisons, all over the world. Their performance and behaviour is well understood. However, because they operate at only about 200 °C, they need a noble metal catalyst, and so, like the PEM fuel cell, are poisoned by any carbon monoxide in the fuel gas. This means that their fuel processing systems are necessarily complex. Since the PAFC has similar characteristics to the PEM fuel cell, and is more commercially advanced than the MCFC and SOFC, this will be described first.

The molten carbonate electrolyte fuel cell (MCFC) has a history that can be traced back at least as far as the 1920s.[1] It operates at temperatures around 650 °C. The main problems with this type of cell relate to the degradation of the cell components over long periods. The MCFC does, however, show great promise for use in combined heat and power systems, and is discussed in detail in Section 6.4.

The solid oxide fuel cell (SOFC) has also been the object of research for many years. Its development can be traced back to the Nernst "Glower" of 1899. Since the SOFC is a solid state device, it has many advantages from the point of view of mechanical simplicity. The SOFC is also very flexible in the way it can be made, and its possible size. It therefore has scope for a wide variety of applications. SOFCs can be made from a range of different materials, with different operating temperatures, from about 650 to 1000 °C. These issues are described in Section 6.5.

However, before we consider the details of the three different types of medium and high temperature fuel cell, we should consider the main features that are *common* to all the cells. Three out of the four main advantages listed on the previous page related to what could be done with the waste heat. It can be used to reform fuels, provide heat, and drive engines. This means that the PAFC, MCFC, and SOFC **can never be considered simply as fuel cells, but they must always be thought of as an integral part of a complete fuel processing and heat generating system.** The wider system issues are largely the same for all three fuel cell types. The common features are considered under four headings as outlined below.

- These medium and high temperature fuel cells will nearly always use a fuel that will need processing. Fuel reforming is a large topic, and is covered in some detail in the chapter that follows, Chapter 7. The basics of how this impinges on fuel cells, and how it is integrated into the fuel cell system, is explained in Section 6.2.1.

- The question of *fuel utilisation* is a problem with all these cells. The fuel will nearly always be a mixture of hydrogen, carbon oxides and other gases. As the fuel passes through the stack the hydrogen will be used, and so its concentration in the mixture will reduce. This reduces the local cell voltage. If all the hydrogen were used up, then in theory the voltage at the exit of the stack would be at zero. The problem of how much of the hydrogen fuel to use in the fuel cell is a tricky one, and is addressed in Section 6.2.2

- The high temperature gases leaving these cells carry large amounts of heat energy. In many cases it is desirable to use turbines to convert this heat energy into further electrical energy. This can be done with all three of the fuel cells considered here. It is explained in Section 6.2.3 how *this combination of fuel cell and heat engine can lead to unsurpassed levels of efficiency,* with each machine compensating for the practical problems of the other.

- Heat from the stack exhaust gases can be used to pre-heat fuel and oxidant using suitable heat exchangers. Best use of heat within high temperature fuel cell systems is

---

[1] Baur and his associates in Switzerland

an important aspect of system design, and chemical engineers often referred to this as "process integration". To ensure high electrical and thermal system efficiencies, the minimising of *exergy* loss is a key element and "pinch technology" is a method that system designers have in their tool box to help with this. Such heat management aspects of system design are covered in Section 6.2.4.

We first consider these features common to all medium and high temperature fuel cells. Note that since these fuel cells do require processed fuel and since use can be made of the exhaust heat, they have been mainly applied to stationary power generation systems. Although we shall see that the SOFC may find application in some mobile applications, the complexity of fuel processing usually rules out the application of high temperature fuel cells for mobile use.

6.2   Common Features

6.2.1   An introduction to fuel reforming

Fuel reforming is described in detail in Chapter 7. Suffice to say at this stage that the production of hydrogen from a hydrocarbon usually involves "steam reforming." In the case of methane the steam reforming reaction may be written as:-

$$CH_4 + H_2O \rightarrow 3H_2 + CO \qquad [6.1]$$

However, the same reaction can be applied to any hydrocarbon C_xH_y, producing hydrogen and carbon monoxide as before.

$$C_xH_y + xH_2O \rightarrow \left(x + \frac{y}{2}\right)H_2 + xCO \qquad [6.2]$$

In most cases, and certainly with natural gas, the steam reforming reactions are *endothermic*. That is, heat needs to be supplied to drive the reaction forward to produce hydrogen. Again, for most fuels, the reforming has to be carried out at relatively high temperatures, usually above about 500°C. For the medium and high temperature fuel cells, heat required by the reforming reactions can be provided, at least in part, from the fuel cell itself in the form of exhaust heat. In the case of the PAFC, the heat from the fuel cell at around 200°C has to be supplemented by burning fresh fuel gas. This has the effect of reducing the efficiency of the overall system, and typically the upper limit of efficiency of the PAFC is around 40-45% (ref. HHV). For the MCFC and SOFC heat is available from the fuel cell exhaust gases at higher temperatures. If all of this heat is used to promote the reforming reactions (especially when reforming is carried out inside the stack) then the efficiency of these fuel cells can be much higher (typically >50% (ref. HHV).

The carbon monoxide produced with hydrogen in steam reforming is a potential problem. This is because it will poison any platinum catalyst, as used in PEMFCs or PAFCs. With these fuel cells the reformate gas must be further processed by means of the "water-gas shift" reaction (usually abbreviated to the "shift" reaction):

$$CO + H_2O \rightarrow CO_2 + H_2 \qquad [6.3]$$

This reaction serves to reduce the CO content of the gas, by converting it into CO_2. As is explained in Section 7.4.9 this process usually needs two stages, at different temperatures, to achieve CO levels low enough for PAFC systems.

A further complication is that fuels such as natural gas nearly always contain small amounts of sulphur or sulphur compounds, which must be removed. Sulphur is a well-known catalyst poison and will also deactivate the electrodes of all types of fuel cell. Therefore, sulphur has to be removed before the fuel gas is passed to the reformer or stack. Desulphurisation is a well established process that is required in many situations, not just for fuel cells. The process is also described in more detail in Chapter 7. At this stage it is sufficient to know that the fuel must be heated to about 350 °C before entering the desulphuriser, which can be considered as a "black box" for the moment.

The total fuel processing system, especially for a PEMFC or a PAFC that requires very low levels of carbon monoxide, is therefore quite a complex system. The reader should look ahead to Figures 7.4 and 7.5 to see what is meant by this. However, the molten carbonate fuel cell (MCFC) and solid oxide fuel cell (SOFC) that we will be considering later operate at temperatures of about 650 °C or more. This is hot enough to allow the basic steam reformation reaction of equation 6.1 above to occur *within the fuel cell stack itself.* Furthermore, the steam needed for the reaction to occur is also present in the fuel cell, because the product water from the hydrogen/oxygen reaction appears at the *anode* – the fuel electrode. Thus there is the opportunity to simplify the design of MCFC or SOFC systems. Steam can be supplied for the reforming by recirculating anode exhaust gas, for example, making the system self-sufficient in water. It is also shown in the sections dedicated to these fuel cells that both of them can use carbon monoxide as a fuel – it reacts with oxygen from the air to produce electric current, just like the hydrogen.

This means that the MCFC and SOFC systems have the potential to be much simpler than PAFC systems. The fuel will still need to be desulphurised, but the fuel processing system is much less complex.

6.2.2   Fuel Utilisation

The question of fuel utilisation arises whenever the hydrogen for a fuel cell is supplied as part of a mixture, or becomes part of a mixture due to internal reforming. As the gas mixture passes through the cell the hydrogen is utilised by the cell, the other components simply pass through unconverted, and the hydrogen concentration falls. If most of the hydrogen gets used, then the fuel gas ends up containing mostly carbon dioxide and steam. The partial pressure of the hydrogen will then be very low. This will considerably lower the cell voltage.

Back in Chapter 2, Section 2.5, we considered the effects of pressure and gas concentration on the "no loss" open circuit voltage of a fuel cell. The very important Nernst equation was presented in many forms. One form, which is useful in this case, relates the open circuit voltage and the partial pressures of hydrogen, oxygen and steam. This was equation 2.8.

$$E = E^0 + \frac{RT}{2F} \ln \left[\frac{P_{H_2} . P_{O_2}^{\frac{1}{2}}}{P_{H_2O}} \right] \qquad [6.4]$$

If we isolate the hydrogen term, and say that the hydrogen partial pressure changes from P_1 to P_2, then the change in voltage is:-

$$\Delta V = \frac{RT}{2F} \ln \left(\frac{P_2}{P_1} \right)$$

In this case the partial pressure of the hydrogen is falling as it is used up, so P_2 is always less than P_1, and thus ΔV will always be negative. An issue with higher temperature fuel cells is the RT term in this equation, which means that the voltage drop will be greater for higher temperature fuel cells. There will also be a voltage drop due to oxygen utilisation from the air, since the partial pressure of the oxygen reduces as it passes through the cell. This is less of a problem – one reason being the ½ term applied to the oxygen partial pressure in equation 6.4.

These issues are shown graphically in Figure 6.1 which shows four graphs of voltage against temperature. The first line shows the voltage of a standard hydrogen fuel cell operating at 1.0 bar, with pure hydrogen and oxygen supplies. The second is for a cell using air at the cathode, and a mixture of 4 parts hydrogen, 1 part carbon dioxide at the anode – simulating the gas mixture that would be supplied by reformed methane. The next two lines show the "Nernst Exit Voltage" for 80% and 90% fuel utilisation, with 50% air utilisation in both cases. Here both the air and fuel are flowing in the same direction (co-flow). We can see that the drop in voltage is significant, and that this is a serious problem. If one part of a cell has a lower voltage, then this will result in reduced efficiency. Alternatively, a reduced current might be the result. As expected, the voltage drop increases with temperature and fuel utilisation.

Sometimes the total voltage drop through a high temperature cell can be reduced by feeding the air and fuel through the cell in opposite directions (counter-flow). This means that the part of the cell with the exiting fuel at least has the highest oxygen partial pressure. The result of this is that the Nernstian voltage drop due to fuel and air utilisation is spread more evenly through the cell. However, fuel and air flow directions are also varied to take account of temperature control, which is sometimes helped with the co-flow arrangements.

A factor that has been ignored in Figure 6.1 is that in the molten carbonate and solid oxide fuel cells, as we shall see, the product steam ends up at the anode – essentially the hydrogen is replaced by steam. So, if the partial pressure of the hydrogen decreases, the steam partial pressure will *increase*. Studying equation 6.4 we see that a rise in the partial pressure of steam will cause a fall in the open circuit voltage. However, the effect of this is difficult to model - some of the steam may be used in internal fuel reforming for example. The situation is liable, in practice, to be worse than Figure 6.1 would indicate.

The important conclusion from this is that, in the case of a reformed fuel containing carbon dioxide, or when internal reforming is applied, all the hydrogen can never be consumed in the fuel cell itself. Some of the hydrogen must pass straight through the cell, to be used later to provide energy to process the fuel or to be burnt to increase the heat energy available for heat engines, as we will be considering in Section 6.2.4 below. The optimum fuel utilisation for any system can only be found by extensive modelling using computer simulations of the entire system. As we have said before, the fuel cell cannot be considered in isolation.

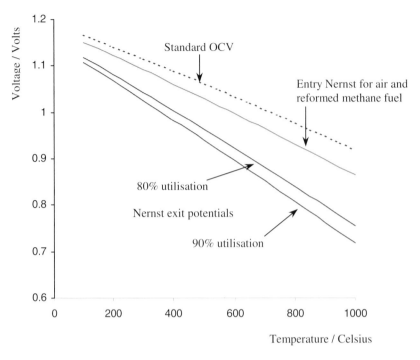

Figure 6.1 Theoretical open circuit voltages for hydrogen fuel cells. The exit graphs refer to reformed methane fuel, and in both cases the oxygen pressure is 0.1 bar, corresponding to 50% air utilisation.

6.2.3   Bottoming cycles

This rather curious term refers to the use of the "waste" heat from a fuel cell exhaust gases to drive some kind of heat engine. There are two ways in which this is commonly done:-

1. The heat is used in a boiler to raise steam, to drive a steam turbine. A diagram of such a system is shown in Figure 6.3.

2. The whole system, including the fuel cell, is pressurised, and the exit gases from the fuel cell power a gas turbine. Such a system is outlined in Section 6.4 on SOFCs, and shown in Figures 6.31 and 6.32.

   A third possibility is to combine both a steam *and* gas turbine with a fuel cell into a triple cycle system. This would be a possibility for larger systems in the future, but at the time of writing none have yet been built.

   These combined cycle systems offer an elegant way of producing very efficient systems of generating electricity. To understand *why* they are potentially so efficient there are two extreme positions that we must recognise and refute. The fuel cell enthusiast might say that the fuel cell acts as a 'reactor', burning the fuel and producing heat to drive the turbine. The electricity is a "free bonus". This is obviously a very efficient system! The opposition might say the fuel cell is a waste. If the fuel was burnt ordinarily, the temperature would be much higher, and so the heat engine would be more efficient. Get rid of the fuel cell!

Both these positions are wrong! To show this we need to use a little thermodynamics theory.[2] In Chapter 2 we saw that the energy accessed in a fuel cell is the Gibbs free energy. We noted, in equation 2.4, that the efficiency limit of a fuel cell was:-

$$= \frac{\Delta \bar{g}_f}{\Delta \bar{h}_f}$$

We can make this equation a little less "cluttered" if we use the plain forms of Gibbs free energy and enthalpy, rather than the molar specific forms. This is justified as long as we do it for both energies, and it gives:-

$$Maximum\, efficiency = \frac{\Delta G}{\Delta H}$$

However, we also noted that in the case of hydrogen the Gibbs free energy *decreases* with temperature. It is the same for carbon monoxide, and so we can say that for all practical fuels this is so. It is impractical to operate a fuel cell at below ambient temperature, and so the highest possible efficiency limit will occur at ambient temperature, T_A. At this temperature the Gibbs free energy is a maximum, and is equal to ΔG_{T_A}. So, in practice, we can refine our formula for the maximum efficiency to:-

$$Maximum\ efficiency = \frac{\Delta G_{T_A}}{\Delta H} \qquad\qquad [6.5]$$

However, at ambient temperature the losses (detailed in Chapter 3) are large, and so the practical efficiencies are well below this limit. To reduce the voltage drops we need to raise the temperature, but this reduces ΔG, and hence the efficiency.

The reason for the fall in ΔG as the temperature rises can be seen from the fundamental thermodynamic relationship that connects the Gibbs free energy, enthalpy, and entropy:-

$$\Delta G = \Delta H - T \Delta S \qquad\qquad [6.6]$$

The enthalpy of reaction ΔH is more or less constant for the hydrogen reaction, as is the entropy change ΔS.[3] The Gibbs free energy becomes less negative (i.e. less energy is released) with temperature because ΔS is negative.

---

[2] The approach given here is along the lines used by A.J.Appleby in Appleby & Foulkes (1993) and Blomen et al. (1993) among others. It does involve some approximations, but what it lacks in complete rigour it more than makes up for in clarity. For an alternative approach based on the T/S diagrams loved (or hated) by students of thermodynamics the reader should consult Gardner (1997).

Now, the fraction of the enthalpy of reaction that is not converted into electricity in the fuel cell is converted into heat. This means that as the temperature of the fuel cell rises, the energy $T \Delta S$ that is converted into heat increases.

Suppose a fuel cell is operating *reversibly*, at a temperature T_F, which is considerably higher than the ambient temperature T_A. Under these reversible conditions:-

- ΔG_{T_F} is the electrical energy that would be supplied by the fuel cell, and

- $T_F \Delta S$ is the heat that would be produced.

Now, this heat energy could be converted into work, and hence electrical energy, using a heat engine and generator. However, even in a reversible system this conversion would be subject to the "Carnot limit". The upper temperature is T_F, the temperature of the fuel cell, as this is the temperature at which the gases leave the cell. The lower temperature is the ambient temperature T_A. So, the extra electrical energy that could be available in a reversible system is:-

$$\frac{T_F - T_A}{T_F} T_F \Delta S = (T_F - T_A)\Delta S$$

This energy would be added to the electrical energy supplied directly by the fuel cell, so the total electrical energy available from the dual cycle system would be given by the equation:-

$$Max.\ Elec.\ Energy = \Delta G_{T_F} + (T_F - T_A)\Delta S$$
$$= \Delta G_{T_F} + T_F \Delta S - T_A \Delta S \qquad [6.7]$$

However, this equation can be simplified, because if we rewrite equation 6.6 specifically for temperature T_F and T_A, we get the following two equations:-

$$T_F \Delta S = \Delta H - \Delta G_{T_F}$$
$$\text{and} \qquad T_A \Delta S = \Delta H - \Delta G_{T_A}$$

If we substitute these two equations into equation 6.7 we get:-

$$Max.\ Elec.\ Energy = \Delta G_{T_F} + \Delta H - \Delta G_{T_F} - \Delta H + \Delta G_{T_A}$$
$$= \Delta G_{T_A}$$

So, dividing this by the enthalpy change to give the fraction of the energy converted into electricity, we have, **exactly as before**:-

$$Maximum\ efficiency = \frac{\Delta G_{T_A}}{\Delta H}$$

---

[3] This can be confirmed by looking at Table A1.3 in Appendix 1, where a range of values are given.

This is a remarkable result! The equation is the same as equation 6.5. If we take a fuel cell and operate it at the elevated temperature of T_F, and we add a heat engine bottoming cycle, we end up with the same efficiency limit as for an ambient temperature fuel cell. Figure 6.2 is a sketch graph of the efficiency limits of a fuel cell on its own, a heat engine on its own, and a combined cycle fuel cell and heat engine. It is hoped that this diagram makes the point – the dual cycle system of fuel cell and heat engine is particularly advantageous.

The elegance of this arrangement becomes even more clear if we consider the limiting case of a heat engine. The Carnot limit is well known, viz.:-

$$Maximum\ Efficiency = \frac{T_2 - T_1}{T_2} \qquad [6.8]$$

However, does this mean we can keep increasing T_2, getting nearing to 100%

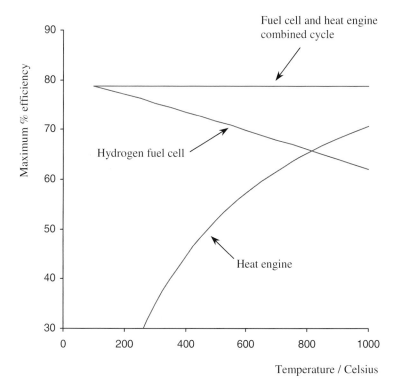

Figure 6.2 Efficiency limits for a heat engine, a hydrogen fuel cell, and a fuel cell/turbine combined cycle. The lower temperature is 100 °C. The fuel cell efficiencies are referred to the higher heating value.

efficiency? No! There is clearly a temperature limit. One of these is certainly the combustion temperature of the fuel being used - it is clear that this cannot reasonably be exceeded. Now, at spontaneous combustion $\Delta G = 0$, and so, using equation 6.4, to a good approximation:-

$$T_c = \frac{\Delta H}{\Delta S}$$

So, we can substitute this equation for T_2 in the Carnot limit equation given above. It is obvious too that the sink temperature T_1 cannot be lower that the ambient temperature T_A, so we can also substitute T_A for T_1. The Carnot limit equation thus becomes:-

$$Maximum \; Efficiency \; = \frac{\dfrac{\Delta H}{\Delta S} - T_A}{\dfrac{\Delta H}{\Delta S}} \; = \; \frac{\Delta H - T_A \, \Delta S}{\Delta H} \qquad [6.9]$$

However, we know that:-

$$\Delta G_{T_A} \; = \; \Delta H - T_A \, \Delta S \qquad [6.10]$$

If we substitute equation 6.10 into 6.9, we see that the maximum efficiency of a fuel burning heat engine is:-

$$Maximum \; Efficiency \; = \; \frac{\Delta G_{T_A}}{\Delta H}$$

This is exactly the same formula again! So we can fully appreciate the potential of the fuel cell/heat engine combined cycle. The heat engine cannot work close to its efficiency limit because of materials and practical limitations. Its efficiency limit is at the combustion temperature of the fuel - but engines cannot operate at temperatures of several thousand Celsius, they would melt! Similarly, though fuel cells can operate at ambient temperatures, their efficiencies are far below the theoretical limit, because of all the problems we discussed in Chapter 3. However, a fuel cell operating at around 800/900/1000 °C can approach the theoretical maximum efficiency. At these temperatures heat engines are also at their best - they do not require exotic materials and are not too expensive to produce. As A.J.Appleby has put it:-

"Thus, a high temperature fuel cell combined with, for example, a steam cycle condensing close to room temperature is a "perfect" thermodynamic engine. The two components of this perfect engine also have the advantage of practically attainable technologies. The thermodynamic losses (i.e. irreversibilities) in a high-temperature fuel cell are low, and a thermal engine can easily be designed to operate at typical heat source temperatures equal to the operating temperature of a high temperature fuel cell. Thus the fuel cell and the thermal engine are complementary devices, and such a combination would be a practical "ideal black box" (or because of its low environmental impact, a "green box") energy system." (Blomen et al., 1993, p.168).

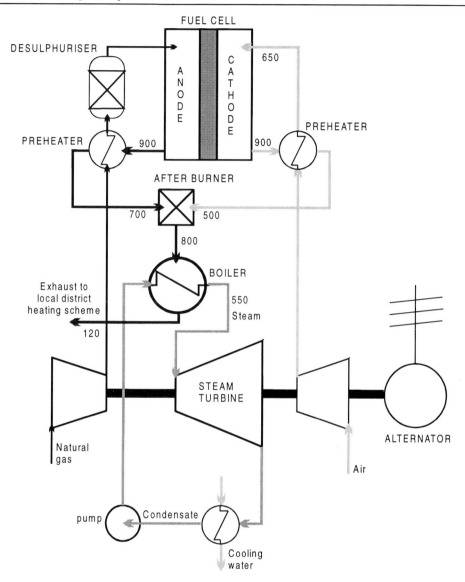

Figure 6.3 Diagram of a possible SOFC and steam turbine combined cycle. The figures indicate likely approximate temperatures of the fluids, in Celsius, at the different stages of the process.

A possible combined cycle system, based on a SOFC, is shown in Figure 6.3. The steam turbine in the centre of the diagram drives two compressors (one for the air, the other for the fuel gas) and an alternator. The air compressor is needed to drive the air through the preheater, the fuel cell, the afterburner, the boiler, and out to the final heating system. The gas compressor needs to drive the fuel through the same components, and the desulphuriser as well, and so would be at a somewhat higher pressure. They would not normally compress the gases much above air pressure, and the fuel cell would operate at about 1.3 bar – only a little above ambient pressure. The natural gas is internally reformed in the fuel cell, but not all the hydrogen is consumed. The remainder is burnt in the afterburner, which raises the temperature of both gas streams to about 800 °C. This hot gas is used in a heat exchanger type boiler to raise steam at about 550 °C, which drives a turbine, which drives the alternator.

The compressors, and the associated work they do driving the gas through the system, represents the major irreversibility in this system. Fry et al. (1997) have published a design of a 36MW system along the lines of that in Figure 6.3, and using realistic figures for turbine and fuel cell performance (as in 1997) and irreversibilities throughout the system, they come up with the following figures:-

Natural gas usage	-	1.19 kg.s^{-1}
Fuel cell output	-	26.9 MWe
Alternator output	-	8.34 MWe

This corresponds to an overall system efficiency of 60%, which is noticeably better than the 55% obtainable from combined cycle gas/steam turbine systems. The combination of fuel cell and gas turbine, where the fuel cell operates at well above ambient pressure, will be presented in the section on SOFCs. Such systems offer the possibility of even higher efficiencies.

6.2.4   The use of heat exchangers – exergy and pinch technology

It will become evident that as more fuel cell systems are described there are challenges in the way in which the various components of the system are integrated. This applies to all systems, but is particularly important in the medium and high temperature systems where several of the balance of plant items are operating at high temperatures. Examples are the desulphuriser, reformer reactor, shift reactors, heat exchangers, recycle compressors and ejectors. In some of these components heat may be generated or consumed. The challenge for the system designer is to arrange the various components in a way that minimises heat losses to the external environment, and at the same time ensures that heat is utilised to the best extent. In this way the energy that is fed to the fuel cell system in the fuel is converted in the most effective way to electricity and useful heat. The optimisation of such system design is known in the chemical engineering discipline as "process integration", and designers will refer to carrying out *exergy analysis* and *pinch technology.*

Heat exchangers
In any fuel cell system there are process streams that need to be heated (e.g. fuel preheat for the reformer, and the reformer itself, and in raising and superheating steam) and streams that need to be cooled (e.g. the fuel cell stack, and the outlet of the shift reactor(s) in a PEM system). The challenge of the system designer is to use the heat available from one stream to heat another in the most efficient way (i.e. avoiding unnecessary heat losses). Heat transfer from one process stream to another is carried out in a *heat exchanger.* The gas (or liquid) to be heated passes through pipework that is heated by the gas (or liquid) to be cooled. A commonly used symbol for a heat exchanger is shown in Figure 6.4. When the exit fluids are used to heat incoming fluids, the heat exchanger is often called a "recuperator."

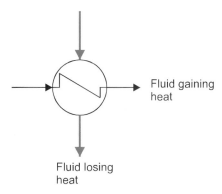

Figure 6.4 A commonly used heat exchanger symbol. The fluid to be heated passes through the "zig-zag" element.

There are several common types of heat exchanger, examples being shell and tube, plate/fin, and printed-circuit exchangers. Descriptions of these are outside the scope of this book, and several good references are available. The choice of heat exchanger is governed by the temperature range of operation, the fluids used (e.g. liquid or gas phase), the fluid throughput and cost. The cost of the exchanger is of course governed by the material costs, but also the heat transfer area. In Chapter 7 we shall see how heat exchangers may be used in practical fuel processing systems. Some are shown being used in Figure 6.3, and Figure 7.4 gives good examples.

Exergy

Exergy may be defined as the quality of heat. More precisely it is work or the ability of energy to be used for work or converted to work. *Potential energy* as classically defined is therefore *exergy*, as is the Gibbs free energy (with the sign changed) for the combustion of fuel. Energy is conserved in all processes (first law of thermodynamics), whereas *exergy* is only conserved in processes that are reversible. Real processes are of course irreversible, so that exergy is always partly consumed to give enthalpy.

The exergy of a system is measured by its displacement from thermodynamic equilibrium. Since $\partial G = V\partial P - S\partial T + \sum \mu_i \, \partial n_i$, where V and P are volume and pressure respectively, and μ_i is the chemical potential of the ith chemical component with a number of molecules, n, it follows that the exergy change of a system going between an initial state and a reference state (subscript o) is given by:

$$E = S(T - T_o) - V(P - P_o) - \sum n_i (\mu_i - \mu_o) \qquad [6.11]$$

The symbol E for exergy used here should not be confused with the same symbol used for energy in classical thermodynamics, nor for fuel cell voltage! It is plain that the higher the temperature T, the greater the exergy. This link with temperature and ability to do work can also be approached less generally via consideration of the Carnot limit.

If we consider the case where both the SOFC and PEM systems have the same power output and efficiency, then the *heat*, that is the enthalpy content, of the exhaust streams produced by both systems will be the same. However, the heat produced by the SOFC will be at a higher temperature and therefore has a higher *exergy* than that from a PEM fuel cell. We can therefore say that the heat from an SOFC is more valuable. The heat that is liberated in a PEM fuel cell is at around 80°C and has limited use both within the system and for external applications. For the latter, it may be limited to air space heating, or possibly integration with an absorption cooling system. The heat that is produced by an SOFC on the other hand, is exhausted at temperatures of around 1000°C and is clearly more valuable, as it can also be used, as we have seen, in a bottoming cycle.

The importance of the concept of exergy is that with high temperature fuel cell systems, the system configuration should be designed in such a way that exergy loss is minimised. If the high grade heat that emerges from the fuel cell stack is used inefficiently within the system then the heat that emerges from the system for practical use will be degraded.

Pinch analysis and system design
Pinch analysis or pinch technology is a methodology that can be applied to fuel cell systems for deciding the optimum arrangement of heat exchangers and other units so as to minimise exergy loss. It was originally designed by chemical engineers as a tool for defining energy saving options, particularly through improved heat exchanger network design. Pinch analysis has been used in many fuel cell system designs, and good examples are given in the reference by Blomen et al. (1993). The concept is straightforward enough but for complex systems, sophisticated computer models are required. The procedure is broadly as follows.

In any fuel cell plant, there will be process streams that need heating (cold streams) and those that need cooling (hot streams) *irrespective of where heat exchangers are located.* The first stage in system design is therefore to establish the basic chemical processing requirements and to produce a system configuration showing all of the process streams, thereby defining the hot and cold streams. By carrying out heat and mass balances, the engineer can calculate the enthalpy content of each process stream. Knowing the required temperatures of each process stream, heating and cooling curves can be produced. Examples of these for an MCFC system are shown in Figure 6.5. The individual cooling and heating curves are then summed together to make two composite curves, one showing the total heating required by all of the process streams and the other showing the total cooling required by the system. These two curves are slid together along the y axis and where they "pinch" together with a minimum temperature difference of e.g. 50 °C, the temperature is noted. This so-called pinch temperature defines the target for optimum process design, since in a real system heat cannot be transferred from above or below this temperature. In some fuel cell systems a pinch temperature is not found, in which case a threshold is defined. Either way, pinch technology provides an excellent method for system optimisation.

Of course, other considerations need to be taken into account in system design. For example the choice of materials for the balance of plant components and their mechanical layout. In the remainder of the book various system configurations are given. These are usually drawn as process flow diagrams (PFD) showing the logical arrangement of the fuel cell stack and the balance of plant components. Many computer models are now available that can be used for calculating the heat and material flows around the system. Temperatures and enthalpies of the process streams can then be calculated and these are usually the first step in system design. What normally follows is a first stage system optimisation, using pinch technology, followed by detailed design and fabrication. Many other techniques are available to the system designer which are outside the scope of this book, and the following descriptions of fuel cell systems will indicate the opportunities for creativity in system design.

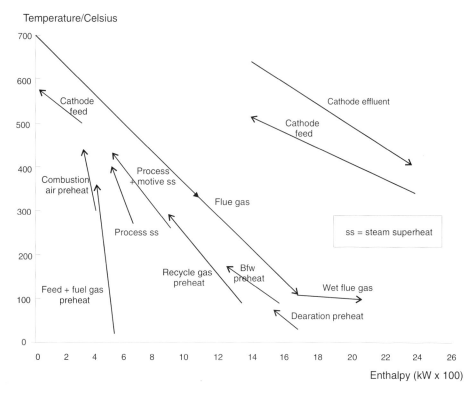

Figure 6.5 Hot and cold heating curves for a conceptual 3.25 MW MCFC system with high pressure steam generation

Having looked at the most important features that are common to all types of "hot" fuel cells, we will now look at the three major types of cell in this category, starting with the Phosphoric Acid Fuel Cell (PAFC).

6.3   The Phosphoric Acid Fuel Cell (PAFC)

6.3.1   How it works

The PAFC works in a similar fashion to the PEM fuel cell described in Chapter 4. The PAFC uses a proton-conducting electrolyte, and the reactions occurring on the anode and cathode are those given in Figure 1.3. In the PAFC, the electrochemical reactions take place on highly dispersed electrocatalyst particles supported on carbon black. As with the PEM fuel cells, platinum (Pt) or Pt alloys are used as the catalyst at both electrodes. The electrolyte is an inorganic acid, concentrated phosphoric acid (100%) which, like the membranes in the PEM cells, will conduct protons.

The electrolyte
Phosphoric acid (H_3PO_4) is the only common inorganic acid that has good enough thermal stability, chemical and electrochemical stability and low enough volatility above about 150°C, to be considered as an electrolyte for fuel cells. Most importantly, phosphoric acid is tolerant to CO_2 in the fuel and oxidant, unlike the alkaline fuel cell. Phosphoric acid was therefore chosen by the US company United Technologies (later the spin-off ONSI Corporation) back in the 1970s as the preferred electrolyte for terrestrial fuel cell power plants.

Phosphoric acid is a colourless viscous hygroscopic liquid. In the PAFC it is contained by capillary action (it has a contact angle >90°) within the pores of a matrix made of silicon carbide particles held together with a small amount of PTFE. The pure 100% phosphoric acid, used in fuel cells since the early 1980s, has a freezing point of 42°C, so to avoid stresses developing due to freezing and re-thawing, PAFC stacks are usually maintained above this temperature once they have been commissioned. Although the vapour pressure is low, some acid is lost during normal fuel cell operation over long periods at high temperature. The loss depends on the operating conditions (especially gas flow velocities and current density). It is therefore necessary to replenish electrolyte during operation, or ensure that sufficient reserve of acid is in the matrix at the start of operation to last the projected lifetime. The SiC matrix comprising particles of about 1 micron, is 0.1-0.2 mm thick, which is thin enough to allow reasonably low ohmic losses (i.e. high cell voltages) whilst having sufficient mechanical strength and the ability to prevent cross-over of reactant gases from one side of the cell to the other. This latter property is a challenge for all liquid based electrolyte fuel cells (see also the MCFC later). Under some conditions, the pressure difference between anode and cathode can rise considerably, depending on the design of the system. The SiC matrix presently used is not robust enough to stand pressure differences greater than 100-200 mbar.

The electrodes and catalysts
Like the PEM fuel cell, the PAFC uses gas diffusion electrodes. In the mid-1960s, the porous electrodes used in the PAFC were PTFE-bonded Pt black, and the loadings were about 9 mg Pt.cm^2 on each electrode. Since then, Pt supported on carbon has replaced Pt black as the electrocatalyst, as for the PEMFC, and as shown in Figure 4.6. The carbon is bonded with PTFE (about 30-50 wt%) to form an electrode support structure. The carbon has important functions:

- To disperse the Pt catalyst to ensure good utilisation of the catalytic metal
- To provide micropores in the electrode for maximum gas diffusion to the catalyst and electrode/electrolyte interface
- To increase the electrical conductivity of the catalyst.

By using carbon to disperse the platinum a dramatic reduction in Pt loading has also been achieved over the last two decades - the loadings are currently about 0.10 mg Pt.cm^2 in the anode and about 0.50 mg Pt.cm^2 in the cathode. The activity of the Pt catalyst depends on the type of catalyst, its crystallite size and specific surface area. Small crystallites and high surface areas generally lead to high catalyst activity. The low Pt loadings that can now be achieved result in part from the small crystallite sizes - down to around 2 nm and high surface areas - up to 100 m^2g^{-1}. (See Figure 1.6 on page 6.)

The PTFE binds the carbon black particles together to form an integral (but porous) structure, which is supported on a porous carbon paper substrate. The carbon paper serves as a structural support for the electrocatalyst layer, as well as acting as the current collector. A typical carbon paper used in PAFCs has an initial porosity of about 90%, which is reduced to about 60% by impregnation with 40 wt% PTFE. This wet proof carbon paper contains macropores of 3 to 50 microns diameter (median pore diameter of about 12.5 microns) and micropores with a median pore diameter of about 3.4 nm for gas permeability. The composite structure consisting of a carbon black/PTFE layer on carbon paper substrate forms a stable, three phase interface in the fuel cell, with electrolyte on one side (electrocatalyst side) and the reactant gas environment on the other side of the carbon paper.

The choice of carbon is important, as is the method of dispersing the platinum, and much know-how is proprietary to the fuel cell developers. Nevertheless, it is known that heat treatment in nitrogen to very high temperatures (1000-2000°C) is found to improve the corrosion resistance of carbons in the PAFC. The development of the catalysts for PAFC fuel cells has been reviewed extensively (e.g. Kordesh, 1979, Appelby, 1984, and Kinoshita, 1988) and lifetimes of 40,000 hours are now expected. However, electrode performance does decay with time. This is due primarily to the sintering (or agglomeration) of Pt catalyst particles and the obstruction of gases through the porous structure caused by electrolyte flooding. During operation, the platinum particles have the tendency to migrate to the surface of the carbon and agglomerate into larger particles, thereby decreasing the active surface area. The rate of this sintering phenomenon depends mainly on the operating temperature. An unusual difficulty is that corrosion of carbon becomes a problem at high cell voltages (above ~0.8 v). For practical applications, low current densities, with cell voltages above 0.8 volts, and hot idling at open circuit potential are therefore best avoided with the PAFC.

The stack

The PAFC stack consists of a repeating arrangement of a ribbed bipolar plate, the anode, electrolyte matrix and cathode. In a similar manner to that described for the PEM cell, the ribbed bipolar plate serves to separate the individual cells and electrically connect them in series, whilst providing the gas supply to the anode and cathode respectively, as shown in Figure 1.8. Several designs for the bipolar plate and ancillary stack components are being used by fuel cell developers, and these aspects are described in detail elsewhere (e.g. Appleby and Foulkes, 1993). A typical PAFC stack may contain 50 or more cells connected in series to obtain the practical voltage level required.

The bipolar plates used in early PAFCs consisted of a single piece of graphite with gas channels machined on either side, as shown in Figure 6.6. Machining channels in graphite is an expensive though feasible method of manufacture. Newer manufacturing methods and designs of bipolar plates are now being used. One approach is to build up the plate in several layers. In "multi-component bipolar plates," a thin impervious carbon plate serves to separate the reactant gases in adjacent cells in the stack, and separate porous plates with ribbed channels are used for directing gas flow. This arrangement is known as the ribbed substrate construction and is shown in Figures 6.7

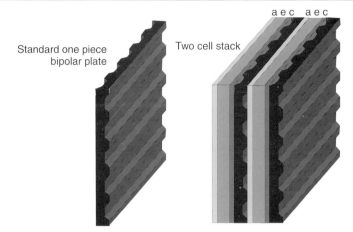

Standard one piece
bipolar plate

Two cell stack

a e c   a e c

Figure 6.6 Standard one piece ribbed bipolar plate

Bipolar plate made from two pieces of porous, ribbed substrate and one sheet
of conductive, impermeable material such as carbon. The porous ribbed
substrates hold phosphoric acid.

aec   aec

Two cell stack.
The electrolyte
can now be thinner,
as the ribbed
substrates also
hold electrolyte.

Figure 6.7 Cell interconnections made from ribbed substrates

The ribbed substrate has some key advantages:

- Flat surfaces between catalyst layer and substrate promote better and uniform gas diffusion to the electrode
- It is amenable to continuous manufacturing process since the ribs on each substrate run in only one direction
- Phosphoric acid can be stored in the substrate, thereby increasing the lifetime of the stack

Stack cooling and manifolding

In PAFC stacks, provisions must be included to remove the heat generated during cell operation. This can be done by either liquid (water/steam or a dielectric fluid) or gas (air) coolants that are routed through cooling channels or pipes located in the cell stack, usually about every fifth cell. Liquid cooling requires complex manifolds and connections, but better heat removal is achieved than with air cooling. The advantage of gas cooling is its simplicity, reliability, and relatively low cost. However, the size of the cell is limited, and the air cooling passages are much larger than those needed for liquid cooling. Water cooling is the most popular method, therefore, and is applied to the ONSI PAFC systems.

Water cooling can be done with either boiling water or pressurised water. Boiling water cooling uses the latent heat of vaporisation of water to remove the heat from the cells. Since the average temperature of the cells is around 180-200°C, this means that the temperature of the cooling water is about 150-180°C. Quite uniform temperatures in the stack can be achieved using boiling water cooling, leading to increased cell efficiency. If pressurised water is used as the cooling medium, the heat is only removed from the stack by the heat capacity of the cooling water, so the cooling is not so efficient as with boiling water. Nevertheless pressurised water gives a better overall performance than using oil (dielectric) cooling or air cooling, though these may be preferred for smaller systems.

The main disadvantage of water cooling is that water treatment is needed to prevent corrosion of cooling pipes and blockages developing in the cooling loops. The water quality required is similar to that used for boiler feed water in conventional thermal power stations. Although not difficult to achieve with ion-exchange resins, such water treatment will add to the capital cost of PAFC systems, and for this reason water cooling is preferred only for units in the 100kW class and above, such as the ONSI PC25 200kW systems.

All PAFC stacks are fitted with manifolds that are usually attached to the outside of the stacks (*external manifolds*). We shall see later that an alternative *internal manifold* arrangement is preferred by some MCFC developers. Inlet and outlet manifolds simply enable fuel gas and oxidant to be fed to each cell of a particular stack. Careful design of inlet fuel manifold enables the fuel gas to be supplied uniformly to each cell. This is beneficial in minimising temperature variations within the stack thereby ensuring long lifetimes. Often a stack is made of several sub-stacks arranged with the plates horizontal, mounted on top of each other with separate fuel supplies to each sub-stack. If the fuel cell stack is to be operated at high pressure, the whole fuel cell assembly has to be located within a pressure vessel filled with nitrogen gas at a pressure slightly above that of the reactants.

6.3.2   Performance of the PAFC

A typical performance (voltage-current) curve for the PAFC is like that shown in Figure 3.1, and follows the usual form for medium to low temperature cells. The operating current density of PAFC stacks is usually in the range 150-400 mA.cm^{-2}. When operating at atmospheric pressure, this gives a cell voltage of between 600 and 800 mV. As with the PEMFC, the major polarisation occurs at the cathode, and like the PEM cell the polarisation is greater with air (typically 560 mV at 300 mA.cm^{-2}) than with pure oxygen (typically 480 mV at 300 mA.cm^{-2}) because of dilution of the reactant. The anode exhibits very low polarisation (-4 mV/100 mA.cm^{-2}) on pure hydrogen, which increases when carbon monoxide is present in the fuel gas. The ohmic loss in PAFCs is also relatively small, amounting to about 12 mV/100 mA.cm^{-2}.

Operating the PAFC at pressure
Cell performance for any fuel cell is a function of pressure, temperature, reactant gas composition and utilisation. It is well known that an increase in the cell operating pressure enhances the performance of all fuel cells, including PAFCs. It was shown in Chapter 2, Section 2.5.4, that for a reversible fuel cell the increase in voltage resulting from a change in system pressure from P_1 to P_2 is given by the formula:-

$$\Delta V = \frac{RT}{4F} \ln\left(\frac{P_2}{P_1}\right)$$

However, this "Nernstian" voltage change is not the only benefit of higher pressure. At the temperature of the PAFC higher pressure also decreases the activation polarisation at the cathode, because of the increased oxygen and product water partial pressures. If the partial pressure of water is allowed to increase, a lower phosphoric acid concentration will result. This will increase ionic conductivity slightly and bring about a higher exchange current density. The important beneficial effect of increasing the exchange current density has been discussed in detail in Section 3.4.2. It results in further reduction to the activation polarisation. The greater conductivity also leads to a reduction in ohmic losses. The result is that the increase in voltage is much higher than this equation would suggest. Quoting experimental data collected over some period, Hischenhofer et al. (1998) suggest the formula:-

$$\Delta V = 63.5 \ln\left(\frac{P_2}{P_1}\right) \tag{6.12}$$

is a reasonable approximation for a temperature range of 177°C < T < 218°C and a pressure range of 1 bar < P < 10 bar.

The effect of temperature
In Chapter 2, Section 2.3, we showed that for a hydrogen fuel cell the reversible voltage decreases as the temperature increases. Over the possible temperature range of the PAFC the effect is a decrease of 0.27 mV/°C under standard conditions (product being water vapour). However, as discussed in Chapter 3, an increase in temperature has a beneficial

effect on cell performance because activation polarisation, mass transfer polarisation, and ohmic losses are reduced. The kinetics for the reduction of oxygen on platinum improve as the cell temperature increases. Again, quoting experimental data collected over some period, Hirschenhofer et al. (1998) suggest that at a mid-range operating load (~250 mA.cm^{-2}), the voltage gain (ΔV_T) with increasing temperature of pure hydrogen and air is given by

$$\Delta V_T = 1.15(T_2 - T_1) \text{ mV} \qquad [6.13]$$

The data suggests that equation 6.13 is reasonably valid for a temperature range of 180°C < T < 250°C. It is apparent from this equation that each degree increase in cell temperature should produce a performance increase of 1.15 mV. Other data indicate that the coefficient for equation 6.13 may actually be in the range of 0.55 to 0.75, rather than 1.15.

Although temperature has only a minimal effect on the H_2 oxidation reaction at the anode, it is important in terms of anode poisoning. Figure 6.8 shows that increasing the cell temperature results in increased anode tolerance to CO poisoning. This increased tolerance is a result of reduced CO adsorption. A strong temperature effect is also seen for simulated coal gas (SCG in Figure 6.8).

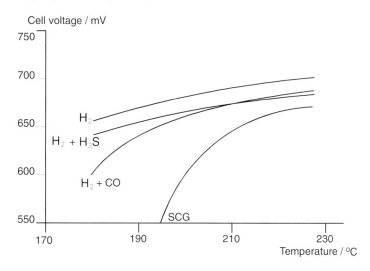

Figure 6.8 Effect of temperature on PAFC cell voltage for different fuels: H_2, H_2 + 200 ppm H_2S and Simulated Coal Gas (Jalan et al., 1990).

Effects of fuel and oxidant utilisation

As mentioned in Section 6.2, fuel and oxidant utilisations are important operating parameters for fuel cells such as the PAFC. In a fuel gas that is obtained, for example by steam reforming of natural gas (see next chapter) the carbon dioxide and unreacted hydrocarbons (e.g. methane) are electrochemically inert and act as diluents. Because the anode reaction is nearly reversible, the fuel composition and hydrogen utilisation generally do not strongly influence cell performance. The RT term in equation 6.4 is clearly lower than for the MCFC and SOFC. Further discussion is given in Hirschenhofer et al. (1998).

On the cathode side, the use of air with ~21% oxygen instead of pure oxygen results in a decrease in the current density of about a factor of three at constant electrode potential. The polarisation at the cathode increases with an increasing oxygen utilisation. Low utilisations therefore, particularly oxygen utilisation, yield high performance. As mentioned in the previous section the drawback of low utilisations is the poor fuel usage and choice of operating utilisations requires a careful balance of all system and stack aspects. State-of-the-art PAFC systems employ utilisations of typically 85% and 50% for the fuel and oxidant respectively.

Effects of carbon monoxide and sulphur

As with the platinum anode catalyst in the PEM fuel cell, the anode of the PAFC may be poisoned by carbon monoxide in the fuel gas. The CO occupies catalyst sites. Such CO is produced by steam reforming and for the PAFC the level that the anode can tolerate is dependent on the temperature of the cell. The higher the temperature, the greater the tolerance for CO. The absorption of CO on the anode electrocatalyst is reversible and CO will be desorbed if the temperature is raised. Any CO has some effect on the PAFC performance, but the effect is not nearly so important as in the PEMFC. At a working temperature above 190°C a CO level of up to 1% is acceptable, but some quote a level of 0.5% as the target. The methods used to reduce the CO levels are discussed in the next chapter, especially Section 7.4.9.

Sulphur in the fuel stream, usually present as H_2S will similarly poison the anode of a PAFC. State of the art PAFC stacks are able to tolerate around 50 ppm of sulphur in the fuel. Sulphur poisoning does not affect the cathode, and poisoned anodes can be re-activated by increasing the temperature or by polarisation at high potentials (i.e., operating cathode potentials).

6.3.3   Recent developments in PAFC

Until recently the PAFC was the only fuel cell technology that could be said to be available commercially. Systems are now available that meet market specifications, and they are supplied with guarantees. An example has already been shown in Figure 1.12, which illustrates the ONSI Corporation 200 kW CHP system. Many of these systems have now run for several years, and so there is a wealth of operating experience from which developers and end users can draw. One important aspect that has come from field trials of the early PAFC plants is the reliability of the stack and the quality of power produced by the systems. The attribute of high power quality and reliability is leading to systems being preferred for so-called premium power applications, such as in banks, in hospitals and in computing facilities. There are now over 65 MW of demonstrators, worldwide, that have been tested, are being tested, or are being fabricated. Most of the plants are in the 50 to 200 kW capacity range, but large plants of 1 MW and 5 MW have also been built. The largest plant operated to date has been that built by International Fuel Cells and Toshiba for Tokyo Electric Power. This has achieved 11 MW of grid quality AC power. Major efforts in the U.S. and Japan are now concentrated on improving PAFCs for stationary dispersed power and on-site cogeneration (CHP) plants. The major industrial developers are the United Technologies spin-off ONSI Corporation (formerly International Fuel Cells) in the U.S. and Fuji Electric, Toshiba, and Mitsubishi Electric Companies in Japan.

Phosphoric acid electrode/electrolyte technology has now reached a level of maturity where developers and users are focusing their resources to producing commercial capacity, multi-unit demonstrations and pre-prototype installations. Cell components are being manufactured at scale and in large quantities. However, the technology is still too costly to be economic compared with alternative power generations systems, except perhaps in niche premium power applications. There is a need to increase the power density of the cells and reduce costs, both of which are inextricably linked. System optimisation is also a key issue. For further information on general technical aspects the interested reader should consult other references e.g. Appelby & Foulkes (1993), Blomen et al. (1993), and Hirschenhofer et al. (1998). Much of the recent technology developments are proprietary but the following gives an indication of progress made during the last few years.

During the early 1990s the goal of the US company IFC was to design and demonstrate a large stack with a power density of 0.188 $W.cm^{-2}$, 40,000 hour useful life, and a stack cost of less than \$400/kW. A conceptual design of an improved technology stack operating at 8.2 atm and 207°C was produced. The stack would be composed of 355 × $1m^2$ cells and produce over 1 MW DC power in the same physical envelope as the 670 kW stack used in the 11 MW PAFC plant built for Tokyo Electric Power. The improvements made to the design were tested in single cells, and in subscale and full size short stacks. The results of these tests were outstanding. The power density goal was exceeded with 0.323 $W.cm^{-2}$ being achieved in single cells operating at 645 $mA.cm^{-2}$ and up to 0.66 volts per cell. In stacks, cell performances of 0.307 $W.cm^{-2}$ have been achieved, with an average of 0.71 volts/cell at 431 $mA.cm^{-2}$. In comparison the Tokyo Electric Power Company's 11 MW power plant, in 1991, had an average cell performance of approximately 0.75 volts per cell at 190 $mA.cm^{-2}$. The performance degradation rate of the advanced stacks was less than 4 mV/1000 hours during a 4500 hour test. The results from this program represent the highest performance of full size phosphoric acid cells and short stacks published to date.

Mitsubishi Electric Corporation have also demonstrated improved performances in single cells of 0.65 mV at 300 $mA.cm^{-2}$. Component improvements by Mitsubishi have resulted in the lowest PAFC degradation rate publicly acknowledged, 2 mV/1000 hours for 10,000 hours at 200 to 250 $mA.cm^{-2}$ in a short stack with 3600 cm^2 area cells.

Catalyst development is still an important aspect of the PAFC. Transition metal (e.g., iron or cobalt) organic macrocycles have been evaluated as cathode electrocatalysts in PAFCs (Bett, 1999). Another approach has been to alloy Pt with transition metals such as Ti , Cr , V, Zr, and Ta. Johnson Matthey Technology Centre have also obtained a 50% improvement in cathode electrocatalyst performance using platinum alloyed with nickel (Buchanan et al., 1992). Work has been done by Giner Inc. and others to make the anode electrocatalysts more tolerant to carbon monoxide and sulphur.

Other recent significant developments in PAFC technology are improvements in gas diffusion electrode construction, and tests on materials that offer better carbon corrosion protection. Of course many improvements can be made in the system design, with better balance-of-plant components such as the reformer, shift reactors, heat exchangers, and burners. Much of this is covered in the chapters that follow. For example, Figures 7.4 and 7.5 show schematic arrangements of the essential components in a PAFC system. In both cases the actual fuel cell stack is a small part of the system.

6.4   The Molten Carbonate Fuel Cell (MCFC)

6.4.1   How it works

The electrolyte of the molten carbonate fuel cell is a molten mixture of alkali metal carbonates – usually a binary mixture of lithium and potassium, or lithium and sodium carbonates, which is retained in a ceramic matrix of $LiAlO_2$. At the high operating temperatures (typically 600-700°C) the alkali carbonates form a highly conductive molten salt, with carbonate CO_3^{2-} ions providing ionic conduction. This is shown schematically in Figure 6.9 which also shows the anode and cathode reactions. Note that unlike all of the common fuel cells, carbon dioxide needs to be supplied to the cathode as well as oxygen, and this becomes converted to carbonate ions, which provide the means of ion transfer between the cathode and the anode. At the anode, the carbonate ions are converted back into CO_2. There is therefore a net transfer of CO_2 from cathode to anode; one mole of CO_2 is transferred along with two Faradays of charge or two moles of electrons. (Note that the requirement for CO_2 to be supplied to the MCFC contrasts with the AFC we considered in the last chapter, where CO_2 must be excluded). The overall reaction of the MCFC is therefore:

$$H_2 + \tfrac{1}{2}O_2 + CO_2 \ (cathode) \ \rightarrow \ H_2O \ + \ CO_2 \ (anode) \qquad [6.14]$$

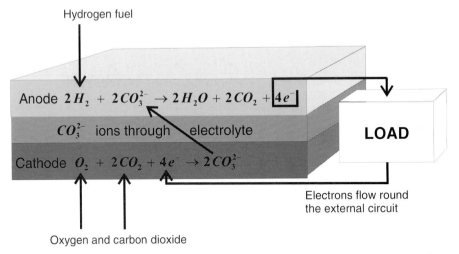

Figure 6.9 The anode and cathode reaction for a molten carbonate fuel cell using hydrogen fuel. Note that the product water is at the anode, and that both carbon dioxide and oxygen need to be supplied to the cathode.

The Nernst reversible potential for an MCFC, taking into account the transfer of CO_2, is given by the equation:

$$E = E^0 + \frac{RT}{2F}\ln\left(\frac{P_{H2}.P_{O2}^{\frac{1}{2}}}{P_{H2O}}\right) + \frac{RT}{2F}\ln\left(\frac{P_{CO2_c}}{P_{CO2_a}}\right) \qquad [6.15]$$

where the sub-subscripts a and c refer to the anode and cathode gas compartments, respectively. When the partial pressures of CO_2 are identical at the anode and cathode, and the electrolyte is invariant, the cell potential depends only on the partial pressures of H_2, O_2, and H_2O. Usually the CO_2 partial pressures are different in the two electrode compartments and the cell potential is therefore affected accordingly, as indicated by equation 6.15.

It is usual practice in an MCFC system that the CO_2 generated at the cell anodes is recycled externally to the cathodes where it is consumed. This might at first seem an added complication, and a disadvantage for this type of cell, but this can be done by feeding the anode exhaust gas to a combustor (burner), which converts any unused hydrogen or fuel gas into water and CO_2. The exhaust gas from the combustor is then mixed with fresh air and fed to the cathode inlet, as is shown in Figure 6.10. This process is no more complex than for other 'hot' fuel cells, as the process also serves to pre-heat the reactant air, burn the unused fuel, and bring the waste heat into one stream for use in a bottoming cycle or for other purposes.

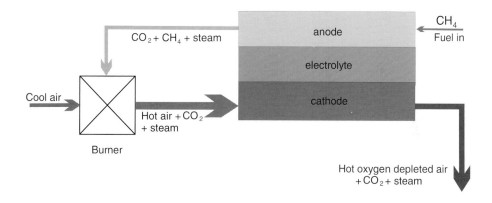

Figure 6.10 Adding CO_2 to the cathode gas stream need not add to the overall system complexity.

Another less commonly applied method is to use some type of device, such as a membrane separator, that will separate the CO_2 from the anode exit gas and allow it to be transferred to the cathode inlet gas (a 'CO_2 transfer device'). The advantage of this method is that any unused fuel gas can be recycled to the anode inlet or used for other purposes. Another alternative to both of these methods is that the CO_2 could be supplied from an external source, and this may be appropriate where a ready supply of CO_2 is available.

At the operating temperatures of MCFCs, nickel (anode) and nickel oxide (cathode) respectively are adequate catalysts to promote the two electrochemical reactions. Unlike the PAFC or the PEMFC, noble metals are not required. Other important differences between the MCFC and the PAFC and PEMFC are the abilities to directly electrochemically convert carbon monoxide and internally reform hydrocarbon fuels. If carbon monoxide were to be fed to the MCFC as fuel, the reactions at each electrode shown in Figure 6.11 would occur.

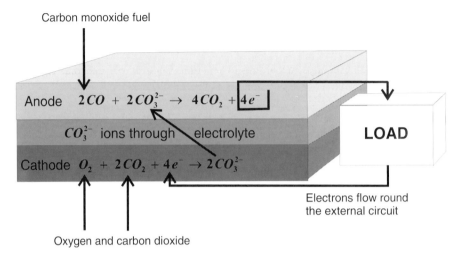

Figure 6.11 The anode and cathode reaction for a molten carbonate fuel cell using carbon monoxide fuel.

The EMF of the carbon monoxide fuel cell is calculated in exactly the same way as for the hydrogen fuel cell, as described in the first section of Chapter 2. Figure 6.10 shows that two electrons are released for each molecule of CO, just as two electrons are released for each molecule of H_2. Thus the formula for the 'no-loss', reversible open circuit voltage is identical, i.e.

$$E = \frac{-\Delta \bar{g}_f}{2F}$$

The method of calculating $\Delta \bar{g}_f$ is given in Appendix 1, and it so happens that at 650°C the values for hydrogen and carbon monoxide are remarkably similar, as is shown in Table 6.1 below.

Table 6.1 Values of $\Delta \bar{g}_f$ and E for hydrogen and carbon monoxide fuel cells at 650 °C.

Fuel	$\Delta \bar{g}_f$ / kJ.mol^{-1}	E / Volts
H_2	-197	1.02
CO	-201	1.04

In practical applications, it is most unlikely that pure CO would be used as fuel. More likely is that the fuel gas would contain both H_2O and CO, and in such cases the electrochemical oxidation of the CO would probably proceed via the water gas shift reaction (equation 6.3), a fast reaction which occurs on the nickel anode electrocatalyst.

The shift reaction converts CO and steam to hydrogen which then oxidises rapidly on the anode. The two reactions (direct oxidation of CO, or shift reaction then the oxidation of H_2), are entirely equivalent

Unlike the PEMFC, AFC, and PAFC, the MCFC operates at a temperature high enough to enable internal reforming to be achieved. This is a particular strong feature of both the MCFC and, as we shall see later, the SOFC. In internal reforming, steam is added to the fuel gas before it enters the stack. Inside the stack, the fuel and steam react in the presence of a suitable catalyst according to reactions such as 6.1 and 6.2. Heat for the endothermic reforming reactions is supplied by the cell electrochemical reactions. Internal reforming is discussed in some detail in the following chapter.

The high operating temperature of MCFCs provides the opportunity for achieving higher overall system efficiencies and greater flexibility in the use of available fuels compared with the low temperature types. Unfortunately, the higher temperatures also place severe demands on the corrosion stability and life of cell components, particularly in the aggressive environment of the molten carbonate electrolyte.

6.4.2   Implications of using a molten carbonate electrolyte

The PAFC and MCFC are similar types of fuel cell in that they both use a liquid electrolyte that is immobilised in a porous matrix. We have seen that in the PAFC, PTFE serves as a binder and wet-proofing agent to maintain the integrity of the electrode structure and to establish a stable electrolyte/gas interface in the porous electrode. The phosphoric acid is retained in a matrix of PTFE and SiC sandwiched between the anode and cathode. There are no materials available that are stable enough for use at MCFC temperatures that are comparable to PTFE. Thus, a different approach is needed to establish a stable electrolyte/gas interface in MCFC porous electrodes. The MCFC relies on a balance in capillary pressures to establish the electrolyte interfacial boundaries in the porous electrodes (Maru and Maianowski, 1976, and Mitteldorf and Wilemski, 1984). This is illustrated in Figure 6.12.

By properly co-ordinating the pore diameters in the electrodes with those of the electrolyte matrix, which contains the smallest pores, the electrolyte distribution shown in Figure 6.12 is established. This arrangement allows the electrolyte matrix to remain completely filled with molten carbonate, while the porous electrodes are partially filled, depending on their pore size distributions. The electrolyte matrix has the smallest pores, and will be completely filled. The larger electrode pores will be partially filled, in inverse proportion to the pore size – the larger the pore, the less they are filled. Electrolyte management, that is the control over the optimum distribution of molten carbonate electrolyte in the different cell components, is critical for achieving high performance and endurance with MCFCs. This feature is very specific to this type of fuel cell. Various undesirable processes (i.e. consumption by corrosion reactions, potential driven migration, creepage of salt and salt vaporisation) occur, all of which contribute to the redistribution of molten carbonate in MCFCs. These aspects are discussed by Maru et al. (1986) and Kunz (1987).

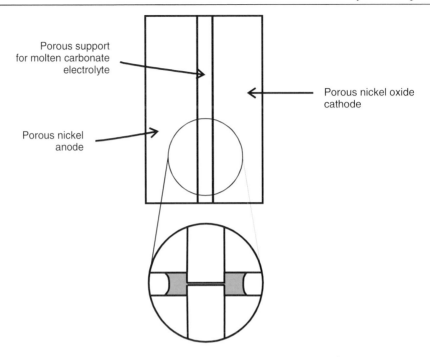

Porous support for molten carbonate electrolyte

Porous nickel oxide cathode

Porous nickel anode

Figure 6.12 Dynamic Equilibrium in Porous MCFC Cell Elements (Porous electrodes are depicted with pores covered by a thin film of electrolyte)

6.4.3   Cell components in the MCFC

In the early days of the MCFC the electrode materials used were, in many cases, precious metals, but the technology soon evolved during the 1960s and 70s to the use of nickel-based alloys at the anode and oxides at the cathode. Since the mid-1970s, the materials for the state-of-the-art electrodes and electrolyte structure (molten carbonate/LiAlO$_2$) have remained essentially unchanged. A major development in the 1980s has been the evolution in the technology for fabrication of electrolyte structures. Several reviews of developments in cell components for MCFCs have been published e.g. by Maru et al. (1984), Petri and Benjamin (1986), and Hirschenhofer et al. (1998). Table 6.2 opposite summarises the evolution of some of the principal materials of the MCFC since the 1960s.

Electrolyte
State-of-the-art MCFC electrolytes contain typically 60wt% carbonate constrained in a matrix of 40wt% LiOAlO$_2$. The γ form of LiOAlO$_2$ is the most stable in the MCFC electrolyte and is used in the form of fibres of <1 μm diameter. Other materials (e.g. larger size particles of LiOAlO$_2$) may be added and many details are proprietary. Until the 1990s the matrix was often fabricated into a tile by a hot pressing process, and it is still often referred to as the electrolyte "tile." Nowadays the matrix is invariably made using tape-casting methods commonly employed in the ceramics and electronics industry. The process involves dispersing the ceramic materials in a "solvent." This contains dissolved

binders (historically organic compounds), plasticisers, and additives to achieve the desired viscosity and rheology of the resulting mixture or "slip". This material is then cast in the form of a thin film over a moving smooth surface, and the required thickness is obtained by shearing with an adjustable blade device. After drying the slip, this material is then heated further in air and any organic binder is burnt out at 250-300°C. The semi-stiff "green" structure is then assembled into the stack structure. Tape casting of the electrolyte and other components provides a means of producing large area components. The methods can also be applied to the cathode and anode materials and fabrication of stacks of electrode area up to 1m^2 is now easily achieved.

The ohmic resistance of the MCFC electrolyte, and especially the ceramic matrix, has an important and large effect on the operating voltage, compared with most other fuel cells. Under typical MCFC operating conditions Yuh et al. (1992) found that the electrolyte accounted for 70% of the ohmic losses. Furthermore, the losses were dependent on the thickness of electrolyte according to the formula:-

Table 6.2 Evolution of Cell Component Technology for Molten Carbonate Fuel Cells (Hirschenhofer et al., 1998).

Component	Ca. 1965	ca. 1975	Current status
Anode	Pt, Pd or Ni	Ni - 10 wt% Cr	Ni-Cr/Ni-Al 3-6 μ m pore size 45-70% initial porosity 0.20-1.5 mm thickness 0.1-1 m^2/g
Cathode	Ag$_2$O or lithiated Nio	Lithiated NiO	lithiated NiO 7-15 μ m pore size 70-80% initial porosity 60-65% after lithiation and oxidation 0.5-1 mm thickness 0.5 m^2/g
Electrolyte Support	MgO	mixture of $\alpha - \beta - \gamma - LiAlO_2$ 10-20 m^2/g	α–LiAlO$_2$, β-LiAlO$_2$ 0.1-12 m^2/g 0.5-1 mm thickness
Electrolyte [a]	52 Li-48 Na 43.5 Li-31.5 Na-25 K "paste"	62 Li-38 K ~60-65 wt% hot press "tile" 1.8 mm thickness	62 Li-38 K 50 Li-50 Na ~50 wt% tape cast 0.5-1 mm thickness

a   Figures in this row are mole percent unless stated otherwise.

$$\Delta V = 0.533t$$

where t is the thickness in cm. From this relationship, it can be seen that a fuel cell with an electrolyte structure of 0.025 cm thickness would operate at a cell voltage 82 mV higher than an identical cell with a structure of 0.18 cm thickness. Using tape-casting methods electrolyte matrices can now be made quite thin (0.25-0.5 mm) which has an advantage in reducing the ohmic resistance. However, there is a trade-off between low resistance and long term stability which is obtained with thicker materials.

Finally, in considering the MCFC electrolyte, we must point out an important difference between the MCFC and all other types of fuel cell. The final preparation of the cell is carried out once the stack components are assembled. So, layers of electrodes, electrolyte and matrix and the various non-porous components (current collectors and bipolar plates) are assembled together, and the whole package is heated slowly up to the fuel cell temperature. As the carbonate reaches its melt temperature (over 450°C) so it becomes absorbed into the ceramic matrix. This absorption process results in a significant shrinkage of the stack components, and provision must be made for this in the mechanical design of the stack assembly. In addition, a reducing gas must be supplied to the anode side of the cell as the package is heated to the operating temperature, to ensure that the nickel anode remains in the reduced state.

Anodes
As indicated in Table 6.2, state of the art MCFC anodes are made of a porous sintered Ni-Cr/Ni-Al alloy. These are usually made with a thickness of 0.4-0.8 mm with a porosity of between 55 and 75%. Fabrication is by hot pressing finely divide powder, or by tape-casting a slurry of the powdered material, which is subsequently sintered. Chromium (usually 10-20%) is added to the basic nickel component to reduce the sintering of the nickel during cell operation. This can be a major problem in the MCFC anode, leading to growth in pore sizes, loss of surface area and mechanical deformation under compressive load in the stack. This can result in performance decay in the MCFC, through re-distribution of carbonate from the electrolyte. Unfortunately, the chromium added to anodes also reacts with lithium from the electrolyte with time, thereby exacerbating the loss of electrolyte. This can be overcome to some extent by the addition of aluminium which improve both creep resistance in the anode, and electrolyte loss. This is believed to be due to the formation of $LiAlO_2$ within the nickel particles. Although Ni-Cr/Ni-Al alloy anodes have achieved commercially acceptable stability, the cost is relatively high and developers are investigating alternative materials. Partial substitution of the nickel with copper, for example, can go some way to reducing the materials costs. Complete substitution of the Ni by Cu is not feasible however, as Cu exhibits more creep than Ni. In an attempt also to improve the tolerance to sulphur in the fuel stream, various ceramic anodes are also being investigated. $LiFeO_2$ with and without dopings of Mn and Nb for example have been tested by Kucerca et al. (1992).

The anode of the MCFC needs to provide more than just electro-catalytic activity. Because the anode reaction is relatively fast at MCFC temperatures, a high surface area is not required, compared with the cathode. Partial flooding of anode with molten carbonate is therefore acceptable, and this is used to good effect to act as a reservoir for carbonate, much in the same way that the porous carbon substrate does in the PAFC. The partial flooding of the anode also provides a means for replenishing carbonate in a stack during prolonged use.

In some earlier MCFC stacks a layer "bubble barrier" was located between the anode and the electrolyte. This consisted of a thin layer of Ni or $LiAlO_2$ containing only small uniform pores. It served to prevent a flow of electrolyte to the anode and at the same time reduced the risk of gas *cross-over*. As we have seen before, this latter problem is common to all liquid fuel cell systems where an excess of pressure on one side of the cell may cause gas to cross the electrolyte. Nowadays, using a tape cast structure it is possible to control the pore distribution of anode materials during manufacture, so that small pores are found closest to the electrolyte and larger pores nearer the fuel gas channels. Long-term electrolyte loss is, however, still a significant problem with the MCFC and a totally satisfactory solution to electrolyte management has yet to be achieved.

Cathodes

One of the major problems with the MCFC is that the nickel oxide state-of-the-art cathode material has a small, but significant, solubility in molten carbonates. Through dissolution, some nickel ions are formed in the electrolyte. These then tend to diffuse into the electrolyte towards the anode. As the nickel ions move towards the chemically reducing conditions at the anode (hydrogen is present from the fuel gas), so metallic nickel can precipitate out in the electrolyte. This precipitation of nickel can cause internal short-circuits of the fuel cell with subsequent loss of power. Furthermore, the precipitated nickel can act as a sink for nickel ions, which promotes the further dissolution of nickel from the cathode. The phenomenon of nickel dissolution becomes worse at high CO_2 partial pressures, because of the reaction:

$$NiO + CO_2 \rightarrow Ni^{2+} + CO_3^{2-} \qquad [6.16]$$

It has been found that this problem is reduced if the more basic, rather than acidic, carbonates are used in the electrolyte. The basicity of the common alkali metal carbonates is: (basic) $Li_2CO_3 > Na_2CO_3 > K_2CO_3$ (acidic). The lowest dissolution rates have been found for the eutectic mixtures 62% Li_2CO_3 + 38% K_2CO_3 and 52% Li_2CO_3 + 48% Na_2CO_3. The addition of some alkaline earth oxides (CaO SrO and BaO) has also been found to be beneficial.

With state of the art nickel oxide cathodes, nickel dissolution can therefore be minimised by (a) using a basic carbonate, (b) operating at atmospheric pressure and keeping the CO_2 partial pressure in the cathode compartment low and (c) using a relatively thick electrolyte matrix to increase the Ni^{2+} diffusion path. By these means, cell lifetimes of 40,000 hours have been demonstrated under atmospheric pressure operation. For operation at higher pressure alternative materials have been investigated, the most studied being $LiCoO_2$ and $LiFeO_2$. Of these two materials $LiCoO_2$ has the lowest dissolution rate, being an order of magnitude lower than NiO at atmospheric pressure. Dissolution of $LiCO_2$ also shows a lower dependency on CO_2 partial pressure than NiO.

Non-porous components

The bipolar plates for the MCFC are usually fabricated from thin sheets of stainless steel. The anode side of the plate is coated with nickel. This is stable in the reducing environment of the anode, it provides a conducting path for current collection and is not wetted by electrolyte which may migrate out of the anode. Gas tight sealing of the cell is achieved by allowing the electrolyte from the matrix to contact the bipolar plate at the edge of each cell outside the electrochemically active area. To avoid corrosion of the stainless steel in this "wet seal" area, the bipolar plate is coated with a thin layer of aluminium. This provides a protective layer of $LiAlO_2$ after reaction of Al with Li_2CO_3. There are many designs of bipolar plate, depending on whether the gases are manifolded externally or internally. Some designs of bipolar plate have been developed especially for internal reforming, such that reforming catalyst can be incorporated within the anode gas flow field. (See Figure 7.3 in the next chapter.)

6.4.4   Stack configuration and sealing

The stack configuration for the MCFC is very different from those described in previous chapters for the PEM, AFC and PAFC, although there are of course similarities. The most important difference is in the aspect of sealing. As described in the previous section, the MCFC stack is composed of various porous components (matrix, and electrodes) and non-porous components (current collector/bipolar plate). In assembling and sealing these components it is important to ensure good flow distribution of gases between individual cells, uniform distribution within each cell, and good thermal management to reduce temperature gradients throughout the stack. Several proprietary designs have been developed for all of the stack components but there are some generic aspects which are described below with examples taken from real systems.

Manifolding
Reactant gases need to be supplied in parallel to all cells in the same stack via common manifolds. Some stack designs rely on external manifolds, whereas others use internal manifolds.

The basic arrangement for *external manifolding* is as shown in Chapter 1, Figure 1.8. The electrodes are about the same area as the bipolar plates, and the reactant gases are fed in and removed from the appropriate faces of the fuel cell stack. One advantage of external manifolding is its simplicity, enabling a low pressure drop in the manifold and good flow distribution between cells. A disadvantage is that the two gas flows are at right angles to each other - cross-flow - and this can cause uneven temperature distribution over the face of the electrodes. Other disadvantages have been gas leakage and migration ("ion-pumping") of electrolyte. Each external manifold must have an insulating gasket to form a seal with the edges of the stack. This is usually zirconia felt, which provides a small amount of elasticity to ensure a good seal. Note that most stack developers arrange the cells to lie horizontal and the fuel and oxidants are supplied to the sides of the stack. An alternative arrangement pioneered by MTU in their "hot module" (see Figure 6.17) is to have vertically mounted cells with the anode inlet manifold located underneath the stack. In this way, sealing with the gasket is enhanced by the weight of the whole stack.

Internal manifolding refers to a means of gas distribution among the cells through channels or ducts within the stack itself, penetrating the separator plates. This is also illustrated in Chapter 1, Figure 1.9. An important advantage of internal manifolding is that there is much more flexibility in the direction of flow of the gases. For even temperature distribution co-flow or counter-flow can be used, as was discussed in Section 6.2.2. The ducts are formed by holes in each separator plate which line up with each other once the stack components are assembled. The separator plates are designed with various internal geometries (flow inserts, corrugations etc.) which as well as providing the walls of the internal manifold, also control the flow distribution across each plate. This is shown in Figure 6.13(c) below. Internal manifolding offers a great deal of flexibility in stack design, particularly with respect to flow configuration. The main disadvantages are associated with the more complex design of the bipolar plate needed.

In internal manifolding, the preferred method of sealing is to use the electrolyte matrix itself as a sealant. The electrolyte matrix forms a wet seal in the manifold areas around the gas ducts. In internally manifolded stacks the entire periphery of the cell may be sealed in this way. It is possible to seal the manifolds with separate gaskets, and extend the matrix only over the active cell area, but this method is rarely used.

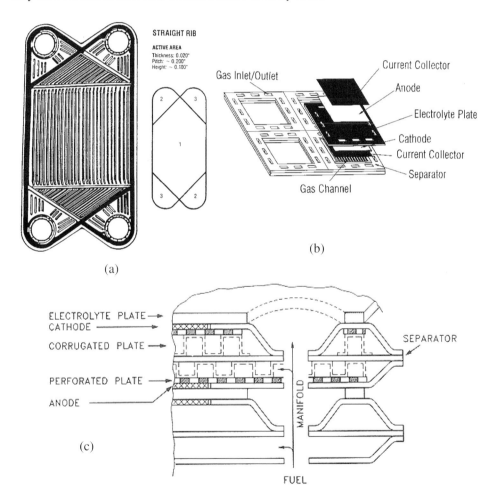

Figure 6.13 Examples of practical separator plate designs with internal manifolding (a) IMHEX design of ECN (b) multiple cell stack of Hitachi. In (c) is shown the cross section of the wet seal area in an internally manifolded MCFC stack.

6.4.5   *Internal reforming*

The concept of internal reforming for the MCFC and SOFC has been mentioned already and will be discussed further in the next chapter. For the MCFC it presents particular challenges. The principal variations of internal reforming configuration are shown in Figure 7.3 in the next chapter. These are described in detail elsewhere (e.g. Dicks, 1998). Direct Internal Reforming (DIR) offers a high cell performance advantage compared with Indirect Internal Reforming (IIR). The reasons for this are quite subtle but it should be noted that with DIR the major product from the reforming reaction, hydrogen, is consumed directly by the electrochemical reaction. Therefore by allowing the reforming reaction and the anode electrochemical reactions to be coupled, the reforming reaction is shifted in the forward direction. *This leads to a higher conversion of hydrocarbon than would normally be expected from reforming equilibrium conditions at the reaction temperature.* At high fuel utilisations in the DIR-MCFC, almost 100% of the methane is converted into hydrogen at 650°C. This compares with a typical conversion of 85%, which would be predicted from the equilibrium of the simple reforming of a steam/methane (2:1 by vol.) mixture at the same temperature. Internal reforming eliminates the cost of an external reformer and system efficiency is improved, but at the expense of potentially a more complex cell configuration. This provides developers of MCFC with a choice of an external reforming or internal reforming approach. Some developers have adopted a combinations of the two methods.

Internal reforming can only be carried out in an MCFC stack if a supported metal catalyst is incorporated. This is because although nickel is a good reforming catalyst, the conventional low surface area porous nickel anode has insufficient catalytic activity in itself to support the steam reforming reaction at the 650°C operating temperature. We shall see that this is not the case for the SOFC, where complete internal reforming may be carried out on the SOFC anode. In the DIR-MCFC the reforming catalyst needs to be close to the anode for the reaction to occur at a sufficient rate. Several groups demonstrated internal reforming in the MCFC in the 1960s and identified the major problem areas to be associated with catalyst degradation, caused by carbon deposition, sintering and catalyst poisoning by alkali from the electrolyte. Most recently, development of internal reforming has been carried out in Europe by BG Technology in a BCN (Dutch Fuel Cell Corporation) led, European Union supported, programme (Kortbeek et al., 1998). Others active in this field have been Haldor Topsoe, working with MTU Friedrichshafen in Germany and Mitsubishi Electric Corporation in Japan. Other companies in the US, Japan and Korea are now involved in internal reforming MCFC technology. Key requirements for MCFC reforming catalysts are as follows.

- **Sustained activity to achieve the desired cell performance and lifetime**. For use in an MCFC, steam reforming catalysts need to provide sufficient activity for the lifetime of the stack so that the rate of the reforming reaction is matched to the rate of the electrochemical reaction, which may decline over a period of time. The strongly endothermic reforming reaction causes a pronounced dip in the temperature profile of an internal reforming cell. This is most severe in the DIR configuration. Optimisation of reforming catalyst activity is therefore important to ensure that such temperature variations are kept to a minimum, to reduce thermal stress, and thereby contribute towards a long stack life. Improvements in temperature distribution across the stack may also be achieved through the recycling of either anode gas, or cathode gas, or both.

- **Resistance to poisons in the fuel.** Most raw hydrocarbon fuels that may be used in MCFC systems (including natural gas) contain impurities (e.g. sulphur compounds) which are harmful for both the MCFC anode and the reforming catalyst. The tolerance of most reforming catalysts to sulphur is very low, typically in the ppb range.

- **Resistance to alkali/carbonate poisoning.** In the case of DIR, where catalysts are located close to the anode there is a risk of catalyst degradation through reaction with carbonate or alkali from the electrolyte. Alkali poisoning and the development of DIR catalysts is described in more detail in the references by Clark et al. (1997) and Dicks (1998), and tolerance to degradation by alkali materials is the biggest challenge for MCFC catalysts. In contrast, many commercial catalysts are available that can be used in IIR-MCFC systems, and it is therefore not surprising that most MCFC developers have adopted the less demanding IIR approach. Commonly, supported nickel catalysts have been used, although more recently researchers have tested supported ruthenium. Poisoning of DIR MCFC reforming catalysts is now known to occur through two principal routes: creep of liquid molten carbonate along the metallic cell components and transport in the gas phase in the form of alkali hydroxyl species. These are illustrated schematically in Figure 6.14, which also indicates one possible solution to the problem, that is the insertion of a protective porous shield between the anode and the catalyst.

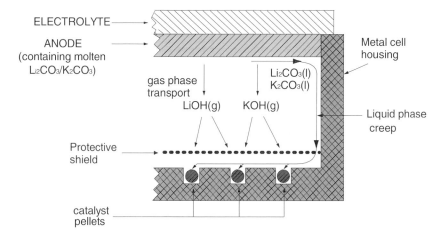

Figure 6.14 Alkali transport mechanisms in the DIR-MCFC

6.4.6   Performance of MCFCs

The operating conditions for an MCFC are selected essentially on the same basis as those for the PAFC. The stack size, efficiency, voltage level, load requirement, and cost are all important and a trade off between these factors is usually sought. The performance (V/I) curve is defined by cell pressure, temperature, gas composition, and utilisation. State-of-the-art MCFCs generally operate in the range of 100 to 200 mA.cm^{-2} at 750 to 900 mV/cell. Figure 6.15 shows the progress made over the last 30 years in improving the performance of generic MCFCs.

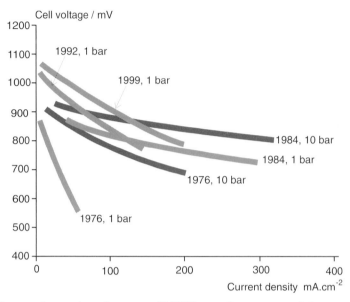

Figure 6.15 Progress in generic performance of MCFCs on reformate gas and air

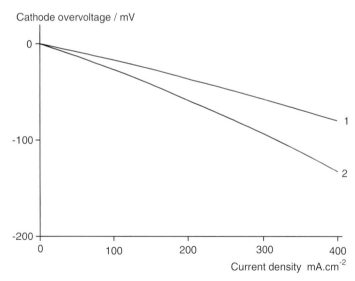

Figure 6.16 Influence of oxidant gas composition on MCFC cathode polarisation at 650°C (1) 33% O_2/67% CO_2 ; (2) 12.6% O_2/18.4% CO_2/69% N_2 (Bregoli and Kunz, 1982)

As with the PAFC, there is significant polarisation of the cathode in the MCFC. This is illustrated in Figure 6.16, which shows typical cathode performance curves obtained at 650°C with an oxidant composition (12.6% O_2/18.4% CO_2/69% N_2) typically used in MCFCs, and a baseline composition (33% O_2/67% CO_2). The baseline composition contains the reactants, O_2 and CO_2, in the stoichiometric ratio that is needed in the electrochemical reaction at the cathode (see Figure 6.9). With this gas composition, little or no diffusion limitations occur in the cathode because the reactants are provided primarily by bulk flow. The other (more realistic) gas composition, which contains a substantial fraction of nitrogen, yields a cathode performance that is limited by gas diffusion and from dilution by an inert gas.

Influence of pressure

There is a performance improvement to be made by increasing the operating pressure of the MCFC. We have already shown in Section 2.5.4 that, from the Nernst equation, for a change in system pressure from P_1 to P_2, the change in reversible potential is given by:

$$\Delta V = \frac{RT}{4F} \ln\left(\frac{P_2}{P_1}\right)$$

From this it can be shown that a fivefold increase in pressure should yield, at 650°C, a gain in OCV of 32 mV, and a tenfold pressure increase should give an increase of 46 mV. In practice the increase is somewhat greater because of the effect of reduced cathode polarisation. Increasing the operating pressure of MCFCs results in enhanced cell voltages because of the increase in the partial pressure of the reactants, increase in gas solubilities, and increase in mass transport rates.

As was shown when considering the PEMFC at pressure in Section 4.7.2, there are important power costs involved in compressing the reactant gases. Also opposing the benefits of increased pressure are the effects on undesirable side reactions such as carbon deposition (Boudouard reaction). Higher pressure also *inhibits* the steam reformation reaction of equation 6.1, which is a disadvantage if internal reforming is being used. These effects, as will be described in the next chapter, can be minimised by increasing the steam content of the fuel stream. In practice the benefits of pressurised operation are significant only up to about 5 bar. Above this there are disadvantages brought about by the system design constraints.

The problem of *differential pressure* is another factor to consider. To reduce the risk of gas cross-over between the anode and the cathode in the MCFC, the difference in pressure between the two sides of each cell should be kept as low as possible. For safety reasons the cathode is usually maintained at a slightly higher (a few mbar) pressure than the anode. The ceramic matrix that constrains the electrolyte in the MCFC is a fragile material that is also susceptible to cracking if subjected to stresses induced either through thermal cycling, temperature variations or excessive pressure differences between the anode and the cathode. The minimisation of pressure difference between anode and cathode compartments in stacks has always been a concern of system designers, since the recycling of anode burner off-gas to the cathode is normally required and there are inevitably pressure losses associated with this transfer of gases. (See Figure 6.10) The problems associated with control of small pressure differences have also mitigated against running the stacks at elevated pressures even though there may be advantages from an efficiency point of view. Since provision of gas tight sealing in the MCFC is achieved through the use of the molten carbonate itself as a sealing medium, this also means that the difference in pressure between the cell compartments and the outside of the stack has to be minimised even when running at elevated pressures. Therefore, any pressurised stack must be enclosed within a pressure vessel in which the stack is surrounded by a non-reactive pressurising gas – usually nitrogen.

Another issue relating to the choice of pressure is that, as we have shown in section 6.3, an improvement in the overall efficiency of fuel systems can be achieved by combining high temperature fuel cells with gas turbines. Gas turbines require pressurised (typically 5 bar) hot exhaust gas. Solid oxide fuel cells are very suitable for this application, as they can run in a pressurised mode and have a high exhaust gas temperature. MCFCs could also be combined with gas turbines but the exhaust temperature is lower. In addition, for the reason described above, the molten carbonate system is not so amenable to pressurisation. Such systems are thus unlikely to be developed.

Generally, it is thought that it is uneconomic to pressurise MCFC systems less than 1MW. More discussion of the influence of pressure is to be found in Hirschenhofer et al.(1998) and in Selman in Blomen et al. (1993).

Influence of temperature
Simple thermodynamic calculations indicate that the reversible potential of MCFCs should decrease with increasing temperature. This is brought about by the Gibbs free energy changes explained in Chapter 2, *and* a change in equilibrium gas composition at the anode with temperature. The main reason for the latter is that the shift reaction (equation 6.3) achieves rapid equilibrium at the anode of the MCFC, and the gas composition depends on the equilibrium for this reaction. The equilibrium constant (K) for the shift reaction increases with temperature, and the equilibrium composition changes with temperature and utilisation to affect the cell voltage, as illustrated in Table 6.3.

Table 6.3 Equilibrium composition for fuel gas and reversible cell potential calculated using the Nernst equation and assuming initial anode gas composition of 77.5% H_2/19.4% CO_2/3.1% H_2O at 1 atmosphere and cathode composition of 30% O_2, 60% CO_2, /10% N_2.

	Temperature (K)		
Parameter	800	900	1000
P_{H2}	0.669	0.649	0.6543
P_{CO2}	0.088	0.068	0.053
P_{CO}	0.106	0.126	0.141
P_{H2O}	0.138	0.157	0.172
E (volts)	1.155	1.143	1.133
K	0.247	0.48	0.711

Under real operating conditions, the influence of temperature is actually dominated by the cathode polarisation. As temperature is increased so the cathode polarisation is considerably reduced. The net effect is that the operating voltage of the MCFC actually increases with temperature. However, above 650 °C this effect is very slight, only about 0.25 mV per °C. Since higher temperatures also increase the rate of all the undesired processes, particularly electrolyte evaporation and material corrosion, 650 °C is generally regarded as an optimum operating temperature (Hirschenhofer et al., 1998).

6.4.7   Practical MCFC systems

Molten carbonate fuel cell technology is being actively developed in the USA, Asia and Europe. An example systems is illustrated below in Figure 6.17.

Figure 6.17 The "hot module" of MTU Friedrichshafen. This unit has an electrical power output of about 250 kW, and features internal reforming. (Pictures reproduced by kind permission of MTU Friedrichshafen.)

Currently, two industrial corporations are actively pursuing the commercialisation of MCFCs in the U.S.: Fuel Cell Energy (formerly Energy Research Corporation) and M-C Power Corporation. Europe and Japan each have at least three developers of the technology: Brandstofel Nederland (BCN), MTU Friedrichshafen (Germany), Ansaldo (Italy), Hitachi, Ishikawajima-Harima Heavy Industries, Mitsubishi Electric Corporation, and Toshiba Corporation. Recently, organisations in Korea have also entered the MCFC field. The technical status of MCFC is exemplified by a 2MW demonstration by Fuel Cell Energy, the "hot module" demonstration by MTU, 250kW demonstrations by MC Power and the 1MW pilot plant currently running in Japan.

A recent review by Dicks and Siddle (1999) indicates that the MCFC is still a few years away from being truly commercial. Nevertheless some important systems have been developed and, by way of example, some *European Systems* are described to provide an indication of their current practical status.

There have been three major industrial developers of MCFC systems in Europe. The Dutch Energy Research Foundation, ECN, brought in other industrial partners to develop an advanced DIR-MCFC system for the European cogeneration market. This consortium developed several advanced system concepts. These included a novel method of stack connection illustrated in Figure 6.18. In this arrangement, shown for 3 stacks but applicable to MCFC systems of 2 or more stacks, stacks are connected together in series on the cathode side, and in parallel on the anode side, with recycle of the anode gas applied. Calculations have shown that the system eliminates the need for all major heat exchangers, and provides a high system efficiency (over 50% HHV).

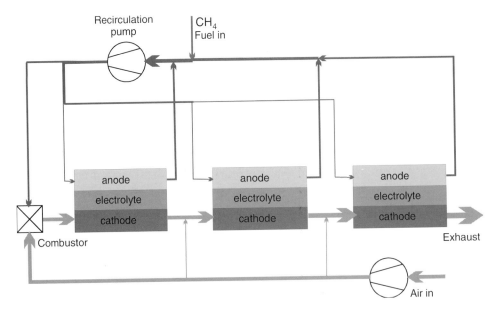

Figure 6.18 Stack networking can simplify system design and increase overall performance

Another concept developed by BG Technology, within the BCN programme, is the so-called SMARTER stack, shown schematically in Figure 6.19. In this stack configuration, two types of cell of slightly different geometries are embodied in the one stack. The two cells differ in that anode recycle is applied to one, and single-pass operation applied to the other. Again, this arrangement shows an efficiency advantage of 5 to 10 percentage points. Interestingly both the series connection and SMARTER stack concepts could be applied to SOFC as well as MCFC systems. The recirculation of the anode gas in these systems has three important advantages.

1. No pre-heating of the fuel supply is needed.
2. The fuel entering the cell has already been partially reformed, and so there is less stress on any direct internal reformation catalyst.
3. The gas flow rates are larger, so there is higher heat capacity in the system, and so the temperature changes through each stack are reduced, giving more even performance.

MTU Friedrichshafen (now part of the DaimlerChrysler company) have developed a "hot module" concept referred to earlier and shown in Figure 6.17. The current status of the system is summarised in Table 6.4 below. Built at a scale of about 260 kW, the system is again designed for industrial and commercial cogeneration applications, such as hospitals. The philosophy behind the design of the "hot module" was as follows:

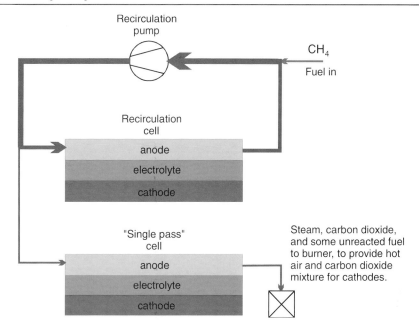

Figure 6.19 BG Technology SMARTER stack concept.

- All hot components are integrated into one common vessel
- Common thermal insulation with internal air recycle for best temperature levelling
- Minimum flow resistance and pressure differences
- Horizontal fuel cell block with internal reforming
- Gravity-sealed fuel manifold
- Simple and elegant mechanics
- Standard truck transportable up to 400 kW and beyond.

Table 6.4 Current status of MTU Friedrichstafen MCFC

Power	279kW (250kW net AC)
No of cells	292
Efficiency	49% LHV
Temperature of available heat	450°C
Stack degradation	1%/1000 hours in a 16000 15x15 cm lab scale test stack
Fuel quality	normal pipeline gas used

For more information on the other developers of MCFC, the review by Dicks and Siddle (1999) should be consulted. Progress is now continuing apace with the leading developers moving from significant demonstrations to early production units. There is a consensus in the international community that commercial systems will be in the market during the first decade of this century. MTU and Fuel Cell Energy are targeting commercial production in 2002. Current demonstrations show that electrical efficiencies of up to 50% (LHV) can be achieved. However, projected electrical and overall efficiencies of 55% and at least 80% are anticipated for the MCFC in the long term. Emissions from these plants are predicted to be negligible with regards to SOx and particulates, and below 10 ppm Nox. Noise levels are below 65dbA. In many cogeneration applications these attributes should give the MCFC an edge over competing technologies such as engine and turbine-based power generation systems.

6.5   The Solid Oxide Fuel Cell

6.5.1.   How it works

The SOFC is a completely solid-state device that uses an oxide ion-conducting ceramic material as the electrolyte. It is therefore simpler in concept than all of the other fuel cell systems described as only two phases (gas and solid) are required. The electrolyte management issues that arise with the PAFC and MCFC do not occur and the high operating temperatures mean that precious metal electrocatalysts are not needed. As with the MCFC, both hydrogen and carbon monoxide can act as fuels in the SOFC, as shown in Figure 6.20[4].

The SOFC is similar to the MCFC in that a negatively charged ion ($O^=$) is transferred from the cathode through the electrolyte to the anode. Thus product water is formed at the anode. Development can be traced back to 1899 when Nernst was the first to describe zirconia[5] (ZrO_2) as an oxygen ion conductor. Until recently SOFCs have all been based on an electrolyte of zirconia stabilised with the addition of a small percentage of yttria (Y_2O_3). Above a temperature of about 800°C, zirconia becomes a conductor of oxygen ions ($O^=$), and typically the state-of-the-art zirconia based SOFC operates between 800 and 1100°C. This is the highest operating temperature of all fuel cells, which presents both challenges for the construction and durability, and also opportunities, for example in combined cycle (bottoming cycle) applications.

The anode of the SOFC is usually a zirconia cermet (an intimate mixture of ceramic and metal). The metallic component is nickel, chosen amongst other things because of its high electronic conductivity and stability under chemically reducing and part-reducing conditions. The presence of nickel can be used to advantage as an internal reforming catalyst, and it is possible to carry out internal reforming in the SOFC directly on the anode (Pointon, 1998). The material for the cathode has been something of a challenge. In the early days of development noble metals were used, but have fallen out of favour on cost grounds. Most SOFC cathodes are now made from electronically conducting oxides or mixed electronically conducting and ion-conducting ceramics. The most common cathode material of the latter type is strontium-doped lanthanum manganite.

---

[4] The high temperatures and presence of steam also means that CO oxidation producing hydrogen, via the shift reaction (equation 6.3), invariably occurs in practical systems, as with the MCFC. The use of the CO may thus be more indirect, but just as useful, as shown in Figure 4.20.
[5] Zirconia is zirconium oxide, yttria is yttrium oxide, etc.

As can be seen from Figure 6.20, unlike the MCFC, the SOFC requires no CO_2 recycling, which leads to system simplification.[6] The absence of CO_2 at the cathode means that the open circuit cell voltage is given by the simple form of the Nernst equation (given in Chapter 2 as 2.8). However, one disadvantage of the SOFC compared with MCFC, is that at the higher operating temperature the Gibbs free energy of formation of water is less negative. This means that the open circuit voltage of the SOFC at 1000°C is about 100 mV lower than the MCFC at 650°C. (See Chapter 2, particularly Figure 2.4 and Table 2.2.) This could lead to lower efficiencies for the SOFC. However, in practice the effect is offset at least in part by the lower internal resistance of the SOFC and the use of thinner electrolytes. The result of this is that the SOFC can be operated at relatively high current densities (up to 1000 mA.cm^{-2}).

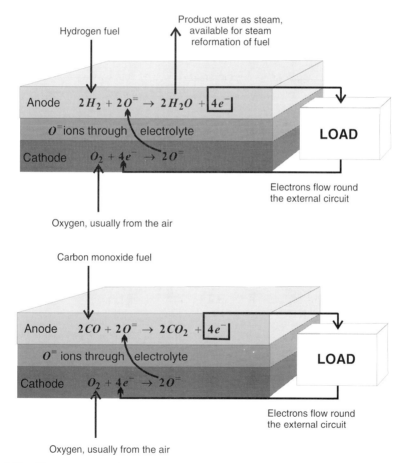

Figure 6.20   The separate anode and cathode reactions for the SOFC, when using hydrogen and carbon monoxide fuel.

---

[6] Though as we noted in Figure 6.10, the simplification might be very slight.

6.5.2   SOFC components

Electrolyte

Zirconia doped with 8-10 mole % yttria (YSZ) is still the most effective electrolyte for the high temperature SOFC although several others have been investigated (Steele, 1994) including Bi_2O_3, CeO_2 and Ta_2O_5. Zirconia is highly stable in both the reducing and oxidising environments which are to be found at the anode and cathode of the fuel cell respectively. The ability to conduct $O^=$ ions is brought about by the fluorite crystal structure of zirconia in which some of the Zr^{4+} ions are replaced with Y^{3+} ions. When this ion exchange occurs, a number of oxide-ion sites become vacant because of three $O^=$ ions replacing four $O^=$ ions. Oxide ion transport occurs between vacancies located at tetrahedral sites in the perovskite lattice. The ionic conductivity of YSZ (0.02 $S.cm^{-1}$ at 800°C and 0.1 $S.cm^{-1}$ at 1000°C) is comparable with that of liquid electrolytes, and it can be made very thin (25-50 μm) ensuring that the ohmic loss in the SOFC is comparable with other fuel cell types. A small amount of alumina may be added to the YSZ to improve its mechanical stability, and tetragonal phase zirconia has also been added to YSZ to strengthen the electrolyte structure so that thinner materials can be made.

Thin electrolyte structures of about 40 μm thickness can be fabricated by "Electrochemical Vapour Deposition" (EVD), as well as by tape casting and other ceramic processing techniques. The EVD process was pioneered by Siemens Westinghouse to produce thin layers of refractory oxides suitable for the electrolyte, anode, and interconnection in these tubular SOFC design (see below). However, it is now only used for fabrication of the electrolyte. In this technique the starting material is a tube of cathode material. The appropriate metal chloride vapour to form the electrolyte is introduced on one side of the tube surface, and an oxygen/steam mixture on the other side. The gas environments on both sides of the tube act to form two galvanic couples. The net result is the formation of a dense and uniform metal oxide layer on the tube in which the deposition rate is controlled by the diffusion rate of ionic species and the concentration of electronic charge carriers.

Zirconia-based electrolytes are suitable for SOFCs because they exhibit pure anionic conductivity. Some materials, such as CeO_2 and Bi_2O_3 show higher oxygen ion conductivities than YSZ but they are less stable at low oxygen partial pressures as found at the anode of the SOFC. This gives rise to defect oxide formation and increased electronic conductivity, which lowers the cell potential (See Section 3.5). Some good progress has been made in recent years in stabilising ceria by the addition of gadolinium (Sahibzada et al., 1997), and adding zinc to lanthanum doped Bi_2O_3. Most recently other materials have been produced with enhanced oxide ion conductivity at temperatures lower than that required by zirconia. The most notable of these is the system LaSrGaMgO (LSGM) (Feng et al., 1996, Ishihara et al., 1994). This material is a superior oxide-ion electrolyte that provides performance at 800 °C comparable to YSZ at 1000 °C, as shown in Figure 6.21.

Anode

The anode of state-of-the-art SOFCs is a cermet made of metallic nickel and a yttria stabilised zirconia skeleton. The zirconia serves to inhibit sintering of the metal particles and provides a thermal expansion coefficient comparable to that of the electrolyte. The anode has a high porosity (20 to 40%) so that mass transport of reactant and product gases is not inhibited. There is some ohmic polarisation loss at the interface between the anode and the electrolyte and several developers are investigating bi-layer anodes in an attempt to reduce this. Most recently the attention of the developers has been directed towards novel ceramic anodes that especially promote the direct oxidation of methane (see Section 7.4.6). Examples are Gd-doped ceria mixed with Zr and Y, and various TiO_2 based systems. In such anodes there is mixed conductivity for both electrons and oxygen ions. A further advantage of using mixed conductors as anodes is that they can provide a means of extending the three-phase boundary between reactant-anode-electrolyte, as shown in Figure 6.22.

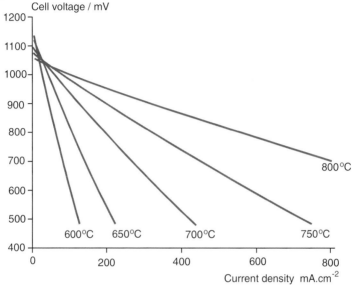

Figure 6.21 Typical single cell performance of LSGM electrolyte (500 m thick)

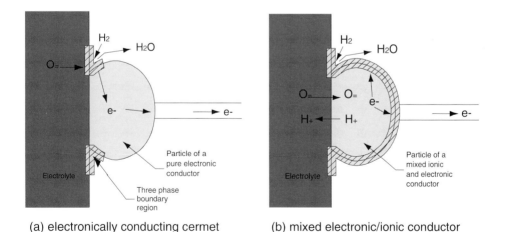

(a) electronically conducting cermet   (b) mixed electronic/ionic conductor

Figure 6.22 Illustration of the three phase boundary regions of different SOFC anode materials. Similar extension of the boundary is obtained in mixed conducting cathode materials.

Cathode

Similar to the anode, the cathode is a porous structure that must allow rapid mass transport of reactant and product gases. Strontium doped lanthanum manganite, a p-type semiconductor, is most commonly used for the cathode material. Although adequate for most SOFCs, other materials may be used, particularly attractive being p-type conducting perovskite structures, which exhibit mixed ionic and electronic conductivity. This is especially important for lower temperature operation since the polarisation of the cathode increases significantly as the SOFC temperature is lowered. It is in cells operating at around 650°C that the advantages of using mixed conducting oxides become apparent. As well as the perovskites, lanthanum strontium ferrite, lanthanum strontium cobalite, and n-type semi-conductors are better electrocatalysts than the state-of-the-art lanthanum strontium manganite, because they are mixed conductors (Han, 1993).

Interconnect material

The "interconnect" is the means by which connection is achieved between neighbouring fuel cells. In planar fuel cell terminology this is the bipolar plate, but the arrangement is different for tubular geometries as will be described in the next section. Metals can be used as the interconnect, but these tend to be expensive "inconel" type stainless steels, particularly for stacks that need to operate at 800-1000°C. An advantage for the low temperature SOFC is that cheaper materials may be used, such as austenitic steels. An alternative, and one that is favoured for the tubular design, is the use of a ceramic material for the interconnect, lanthanum chromite being the preferred choice. The electronic conductivity of this material is enhanced when some of the lanthanum is substituted by magnesium or other alkaline earth elements. Unfortunately, the material needs to be sintered to quite high temperatures (1625°C) to produce a dense phase. This exposes one of the major problems with the SOFC, that of fabrication. All of the cell components need to be compatible with respect to chemical stability and mechanical compliance (similar thermal expansion coefficients). The various layers need also to be deposited in such a way that good adherence is achieved without degrading the material due to the use of too high a sintering temperature. Many of the methods of fabrication are proprietary and considerable research is being carried out in this field.

6.5.3   Design and stacking arrangements for the SOFC

Tubular design

The tubular SOFC was pioneered by the US Westinghouse Electric Corporation (now Siemens-Westinghouse) in the late 1970s. The original design used a porous calcia-stabilised zirconia support tube, 1-2mm thick onto which the cylindrical anodes were deposited. By a process of masking, the electrolyte, interconnect and finally the fuel electrode were deposited on top of the anode. The process was reversed in the early 1980s so that the air electrode became the first layer to be deposited on the zirconia tube, and the fuel electrode was on the outside. This tube became the norm for the next 15 years. Most recently the zirconia support tube has been eliminated from the design and the tubes are now made from air-electrode materials, resulting in an air-electrode-support (AES) onto which the electrolyte is deposited by EVD, followed by plasma spraying of the anode.

Further details of the fabrication methods are given elsewhere, (e.g. Murugesamoorthi et al. in Blomen et al., 1993) and Hirschenhofer et al., (1998). The arrangement of the AES cell is shown in Figure 6.23. The present generation of Siemens-Westinghouse SOFC tubes are 150 cm long and 2.2 cm diameter. They are arranged in series/parallel stacks of 24 tubes as shown in Figure 6.25. Fuel is supplied to the outside of the tubes.

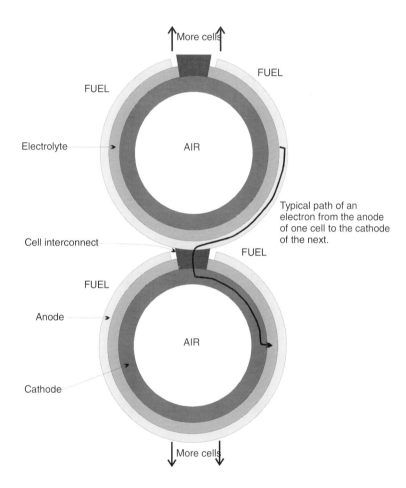

Figure 6.23   End view of tubular type solid oxide fuel cell produced by Siemens-Westinghouse. The electrolyte and the anode are built onto the air cathode.

One great advantage of the tubular design of SOFC is that high temperature gas-tight seals are eliminated. The way this is done is shown in Figure 6.24. Each tube is fabricated like a large test-tube, sealed at one end. Fuel flows along the outside of the tube, towards the open end. Air is fed through a thin alumina air supply tube located centrally inside each tubular fuel cell. Heat generated within the cell brings the air up to the operating temperature. The air then flows through the fuel cell back up to the open end. At this point air and unutilised fuel from the anode exhaust mix are instantly combusted and so the cell exit is above 1000°C. This combustion provides additional heat to preheat the air supply tube. Thus the tubular SOFC has a built in air preheat and anode exhaust gas combustor, as well as requiring no high temperature seals. Finally, by allowing imperfect sealing around the tubes, some recirculation of anode product gas occurs allowing internal reforming (the anode product contains steam and CO_2) of fuel gas on the SOFC anode.

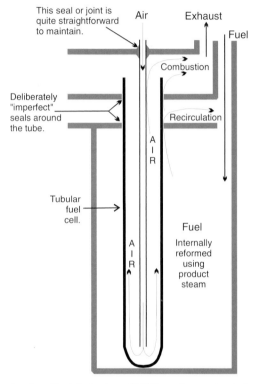

Figure 6.24 Diagram showing how the tubular type SOFC can be constructed with (almost) no seals.

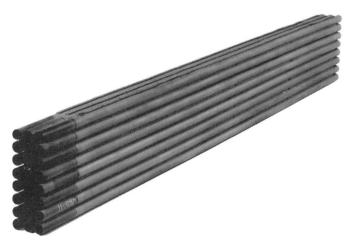

Figure 6.25 Small stack of 24 tubular SOFCs. Each tube has a diameter of 2.2 cm and is about 150 cm long. (Photograph reproduced by kind permission of Siemens Westinghouse.)

Figure 6.26 Larger stack made from bundles of 24 SOFC tubes. There are 1152 cells, and this stack has a power of about 200 kW. (Photograph reproduced by permission of Siemens Westinghouse)

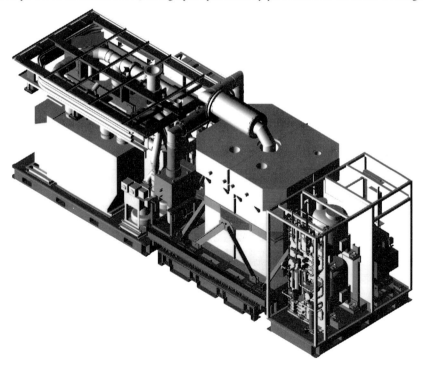

Figure 6.27 Part of a 100 kW SOFC combined heat and power unit. The fuel cell stack is the central unit. The fuel processing system is on the small skid at the front, and the thermal management /electrical/control system is at the rear. The unit is about $8.5 \times 3 \times 3$ m. (Diagram reproduced by kind permission of Siemens Westinghouse.)

Several other organisations, notably Mitsubishi Heavy Industries in Japan and also the group at Keele University in the UK, have been developing tubular SOFC designs. Nevertheless, the Siemens Westinghouse design is the most advanced. Figures 6.25 and 6.26 show stacks built up of tubular fuel cells, and Figure 6.27 shows a 100 kW demonstration which has been built and operated in the Netherlands.

Planar design

Alternatives to the tubular SOFC have been developed for several years, notably several types of planar configuration, and a monolithic design. The planar configurations more closely resemble the stacking arrangements described for the PAFC and PEMFC. This bipolar or flat plate structure enables a simple series electrical connection between cells without the long current path through the tubular cell shown in Figure 6.23. The bipolar flat plate design thus results in lower ohmic losses than in the tubular arrangement. This leads to a superior stack performance and a much higher power density. Another advantage of the planar design is that low-cost fabrication methods such as screen-printing and tape casting can be used.

One of the major disadvantages of the planar design is the need for gas-tight sealing around the edge of the cell components. Using compressive seals this is difficult to achieve and glass ceramics have been developed in an attempt to improve high temperature sealing. Similarly, thermal stresses at the interfaces between the different cell and stack materials tends to cause mechanical degradation, so thermal robustness is important. Particularly challenging is the brittleness of planar SOFCs in tension. The tensile strength of zirconia SOFCs is only about 20% of their compressive strength. Thermal cycling is also a problem for the planar SOFC. Finally the issue of thermal stresses and fabrication of very thin components places a major constraint on the scale up of planar SOFCs. Until recently planar SOFCs could be manufactured only in sizes up to 5×5 cm. Now 10×10 cm planar cells are routinely made (Huijsmans et al., 1999). Such cells may be assembled into a stack by building them into a window-frame arrangement such that several cells are located in one stack layer.

Despite their fundamental attraction in terms of power density and efficiency compared with tubular designs, several organisations have abandoned development of the planar SOFC design because of all of the inherent technical problems. These have included Siemens in Europe who had built 20 kW stacks by 1999. However, planar technology is being carried forward by companies such as Ceramic Fuel Cells Ltd. in Australia, Sulzer Hexis and ECN in Europe, Allied Signal and SOFCo in the US, Rolls Royce in the UK and by companies such as Murata Manufacturing in Japan.

Ceramic Fuel Cells Ltd. have progressed to building 25 kW stacks, and are planning demonstrations at the 100kW scale (Foger et al., 1999). Sulzer Hexis are targeting the small scale cogeneration market with their unique planar stack arrangement (Figure 6.28). In this design, pipeline natural gas is desulphurised and fed to a small pre-reformer or partial oxidation unit located beneath the stack. Future designs will incorporate pre-reforming within the stack. The fuel gas is then fed into the centre of a cylindrical stack made of layers of circular SOFCs interspersed with air manifolds. As shown in Figure 6.28, the design of the bipolar plate ensures that the reactant air is pre-heated. Unreacted fuel is burnt to heat water for the domestic heating system.

Allied Signal, SOFCo, Ztek, Rolls Royce and the various Japanese developers of SOFC are all at the laboratory stage with stacks of up to about 1kW being demonstrated.

The Allied Signal SOFC is a flat-plate concept that involves stacking high-performance thin-electrolyte cells with lightweight metallic interconnect assemblies. Each cell comprises a relatively thick anode which supports a thin film electrolyte. This SOFC design can be operated at temperatures between 600 and 800°C. Single cells used in this design contain supported thin electrolytes. Numerous multicell stacks have been assembled and performance tested, demonstrating power densities of over 650 mW/cm^2 at 800°C. Allied Signal are targeting small scale applications for their technology (Montgomery et al., 1998). Similarly SOFCo has demonstrated several sub kW scale stacks at an operating temperature of 850-900°C with performance degradations of 0.5% per 1000 hours. Zetek have built 1 kW stacks and have a concept for a 100kW class system of 25 kW modules. Rolls Royce have developed a variation on the planar design which they term "Integrated Planar SOFC". This consists of an assembly of small planar SOFCs fabricated on a ceramic housing. The housing serves as a manifold for the fuel gas, with a novel sealing arrangement. Rather than using a bipolar plate, the cells are connected by an interconnector fabricated onto the cell housing.

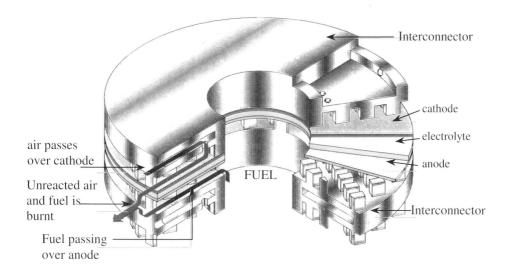

Figure 6.28 Ring type solid oxide fuel cell, with metal cell interconnects. (Diagram reproduced by kind permission of Sulzer Hexis Ltd.)

6.5.4 SOFC performance

As mentioned earlier, with hydrogen as the fuel, the open circuit voltage of SOFCs is lower than that of MCFCs and PAFCs (see discussion in Section 2). However, the higher operating temperature of SOFCs reduces polarisation at the cathode. So, the voltage losses in SOFCs are governed mainly by ohmic losses in the cell components, including those associated with current collection. The contribution to ohmic polarisation in a tubular cell is typically some 45% from cathode, 18% from the anode, 12% from the electrolyte, and 25% from the interconnect, when these components have thickness of 2.2, 0.1, 0.04 and

0.085mm, respectively, and resistivities at 1000°C of 0.013, 3×10^6, 10, and 1 ohm.cm, respectively. The cathode ohmic loss dominates despite the higher resistivities of the electrolyte and cell interconnection because of the short conduction path through these components and the long current path in the plane of the cathode - see Figure 6.23. As we have seen in the previous section SOFCs are being developed with various materials and designs and are being made by several different methods. SOFCs, especially the planar type, may well develop considerably over the next few years and the performance characteristics are likely to change. It could be that the development will lead to a standardisation of cell types, as has been achieved already with PAFC and MCFC. On the other hand several quite different types could come into use, as with the PEMFC. The following gives general indications of the effect of pressure and temperature on SOFC performance. For further discussion, the reader is referred to Hirschenhofer et al. (1998) and the proceedings of the European SOFC Forums.

Influence of pressure
SOFCs, like all fuel cell types, show an enhanced performance with increasing cell pressure. Unlike low and medium temperature cells the improvement is mainly due to the increase in the Nernst potential. We showed in Section 2.5.4 that the voltage change for an increase in pressure from P_1 to P_2 follows very closely the theoretically equation:-

$$\Delta V = 0.027 \ln\left(\frac{P_2}{P_1}\right)$$

The relationship is borne out in practice and Siemens Westinghouse in conjunction with Ontario Hydro Technologies has tested AES cells at pressures up to 15 atmospheres on both hydrogen and natural gas (Singhal, 1997). Operation at higher pressure is particularly advantageous when using the SOFC in a combined cycle system with a gas turbine. In other cases, as with the PEMFC, the power costs involved in compressing the reactants make the benefits marginal.

Influence of temperature
The temperature of an SOFC has a very marked affect on its performance, though the details will vary greatly between cell types and materials used. The predominant effect is that higher temperatures increases the conductivity of the materials, and this reduces the ohmic losses within the cell. As we saw in Chapter 3, ohmic losses are the most important type of loss in the SOFC.

For SOFC-combined cycle and hybrid systems it is beneficial to keep the operating temperature of the SOFC high. For other applications, such as cogeneration, and possible transport applications (the SOFC is being developed by BMW for use as an auxiliary power supply for vehicles), it is an advantage to operate at lower temperatures, as the higher temperatures bring material and construction difficulties. Unfortunately, as Figure 6.21 clearly shows, the performance decreases substantially for SOFCs as the temperature is lowered. Indeed, if zirconia (the standard electrolyte material) is used in place of the LGSM electrolyte of Figure 6.21, the performance falls off at even higher temperatures, and operation at about 900 to 1000°C is required.

As was mentioned in the section on cell interconnects, one of the main advantages of operating at lower temperatures is the possibility of using cheaper construction materials and methods. Making electrolytes and electrodes that work well at lower temperatures is a major focus of current SOFC research.

6.5.5   SOFC combined cycles, novel system designs and hybrid systems

In Section 6.2.3 we discussed how a high temperature fuel cell could be combined with a steam turbine in a bottoming cycle. The ability to use both gas turbines and steam turbines in a combined cycle with an SOFC has been known in concept for many years. However, it is only recently that pressurised operation of SOFC stacks has been demonstrated for prolonged periods, making the SOFC/gas turbine combined cycle system feasible practically. Pioneered by Siemens Westinghouse in their SureCell™ concept, the ideas of combined SOFC/GT are now being explored by other developers. Figure 6.29 shows the design of the 1MW Siemens Westinghouse concept, and the essential process features are shown in Figure 6.30. These systems, and variations on them, are further described in the literature. (For example Veyo & Forbes, 1998, Bevc, 1977, and Fry et al., 1977.)

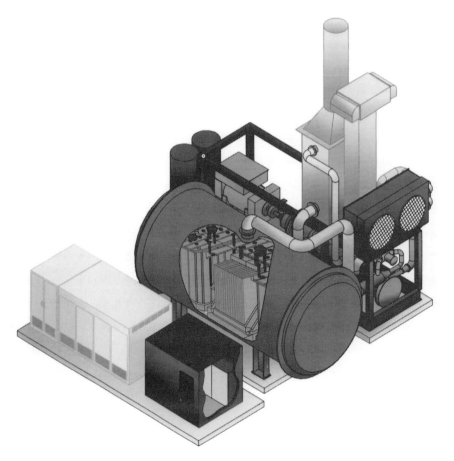

Figure 6.29 Design of a 1 MW SOFC/GT combined cycle plant. The SOFC is shown in the middle of the diagram, and operates at about 10 bar inside the cylindrical pressure vessel. The gas turbine, compressor and alternator are behind the fuel cell.

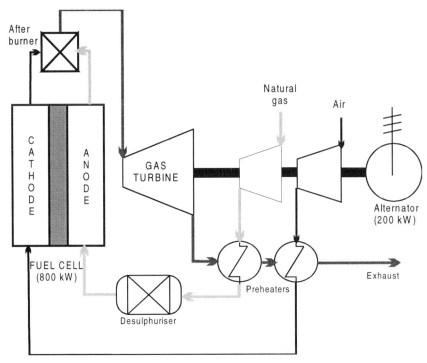

Figure 6.30 System diagram for the SOFC/GT combined cycle system shown in Figure 6.29

In this final section on high temperature fuel cells, it is worthwhile remarking that there are many opportunities for novel system design, and there is scope for considerable creativity by the systems engineer. There are many examples in the literature. For instance, by using the concept of series stack connection mentioned first in connection with the MCFC (section 6.4.6), a "multi-stage" SOFC concept "UltraFuelCell" has been developed by the US Department of Energy and described by Hirschenhofer (1998). However, perhaps one of the most novel ideas, is the hybrid or combined system concept described recently by Siemens-Westinghouse (Vollmar et al., 1999) and BG Technology (Mescal, 1999).

This hybrid system is one in which both SOFC and PEM fuel cells are combined. The advantages of each type of fuel cell are enhanced by operating in synergy. The system shown in Figure 6.31 works as follows. The internal reforming SOFC is run under conditions that give low fuel utilisation. This enables a high power output, for a relatively low stack size. Unutilised fuel appears in the anode exhaust where it undergoes shift reaction, followed by a process stage when the final traces of carbon monoxide are removed. At this stage the gas comprises mainly hydrogen and CO_2, with some steam. This gas, once it is cooled, is suitable for use as a fuel in the PEM stack. The use of two stacks of different types for power generation results in a high overall electrical efficiency. The system becomes particularly attractive if an economic analysis is carried out. Preliminary calculations show that the system is more cost effective than an SOFC-only system because of the anticipated relatively low cost of the PEM stack. On the other hand the system has a much higher efficiency than a natural gas fuelled PEM-only system could achieve. We also see in the following chapter, that the fuel processing technology for running a PEMFC from natural gas is complex, bulky and expensive. How much better to use a SOFC as the fuel processor!

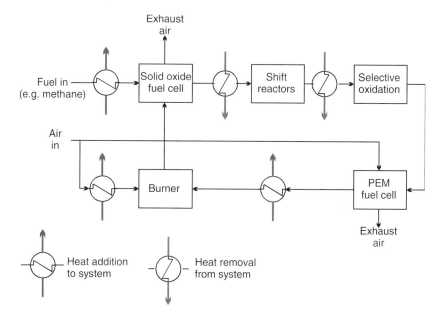

Figure 6.31 SOFC-PEM hybrid system   (Mescal, 1999)

Table 6.5 Summary of output powers for the hybrid system shown in Figure 6.31

SOFC stack power	369.3 kW
PEM stack power	146.7 kW
Turbine power	100.3 kW
Compressor power	-100.8 kW
Net power output	515.5 kW
Electrical output	489.7 kW
Overall efficiency (%) LHV	61

References

Appleby J. (1984) in Proceedings of the Workshop on the Electrochemistry of Carbon, Edited by S. Sarangapani, J.R. Akridge and B. Schumm, The Electrochemical Society, Inc., Pennington, NJ, p. 251.

Appleby A.J. & Foulkes F.R. (1993) *A Fuel Cell Handbook,* 2nd Edition, Krieger Publishing Co.

Badwal S.P.S. & Foger K. (1998) "SOFC development at Ceramic Fuel Cell Limited", Proceedings of the third European Solid Oxide Fuel Cell Forum, pp95-104

Bett J.A.S., Kunz H.R., Smith S.W., and Van Dine L.L., (1985) "Investigation of Alloy Catalysts and Redox Catalysts for Phosphoric Acid Electrochemical Systems," FCR-7157F, prepared by International Fuel Cells for Los Alamos National Laboratory under Contract No. 9-X13-D6271-1

Bevc F. (1997) "Advances in solid oxide fuel cells and integrated power plants", *Proceedings of the Institution of Mechanical Engineers* (London), vol.211, part A, pp359-366

Blomen, Leo J.M.J., & Mugerwa, (1993) *Fuel Cell Systems,*Plenum Publishing

Bosio B., Costamagna P., Parodi F., Passalacqua B. (1998) "Industrial experience on the development of the molten carbonate fuel cell technology", *Journal of Power Sources*, vol.74, no.2, pp175-187

Bregoli L.J. & Kunz H.R.(1982) "The effect of thickness on the performance of molten carbonate fuel cell cathodes", *Journal of the Electrochemical Society,* Vol 129, No. 12, pp 2711-2715.

Brenscheidt T., Janowitz K., Salge H.J., Wendt H., Brammer F. (1998) "Performance of ONSI PC25 PAFC cogeneration plant" *International Journal of Hydrogen Energy*, Vol.23, No1, pp53-56

Buchanan J.S., Hards G.A., Keck L., Potter R.J., (1992) "Investigation into the Superior Oxygen Reduction Activity of Platinum Alloy Phosphoric Acid Fuel Cell Catalysts," in *Fuel Cell Seminar Abstracts*, Tucson, Arizona, U.S., November 29-December 2.

Clarke S.H., Dicks A.L., Pointon, K, Smith T.A., Swann A., (1997) "Catalytic aspects of the steam reforming of hydrocarbons in internal reforming fuel cells." *Catalysis Today* vol 38, pp 411-423.

Dicks A.L. (1998) "Advances in catalysts for internal reforming in high temperature fuel cells." Journal of Power Sources." *Journal of Power Sources* vol. 71, pp 111-122.

Dicks A.L., Siddle A. (1999) "Assessment of Commercial Prospects of Molten carbonate Fuel cells." ETSU Report No.F/03/00168/REP, AEA Technology, Harwell UK

Feng M., Goodenough, J., Huang, K., and Milliken C.,(1996) *Journal of Power Sources* vol. 64 p 47.

Foger K., Godfrey B., and Pham. T., (1999) "Development of 25kW SOFC system" Fuel Cells Bulletin No. 5 February, Elsevier Science Ltd., ISSN 1350-4789, pp. 9-11.

Fry M.R., Watson H., & Hatchman J.C. (1997) "Design of a prototype fuel cell/composite cycle power station", *Proceedings of the Institution of Mechanical Engineers* (London), vol.211, part A, pp171-180

Gardner F.J. (1997) "Thermodynamic processes in solid oxide and other fuel cells", *Proceedings of the Institution of Mechanical Engineers* (London), vol.211, part A, pp367-380

Han P., et al., (1993) "Novel Oxide Fuel Cells Operating at 600 - 800°C," An EPRI/GRI Fuel Cell Workshop on Fuel Cell Technology Research and Development, New Orleans, LA, April 13-14.

Hirschenhofer J., (1993) "Status of Fuel Cell Commercialization Efforts," *American Power Conference*, Chicago, Illinois, U.S. April.

Hirschenhofer J.H., Stauffer D.B., Engelman R.R., and K;lett M.G., (1998*) Fuel Cell Handbook Fourth Edition*, Parsons Coporation, for U.S. Department of Energy Report no. DOE/FETC-99/1076.

Huijsmans J.P.P., Huiberts R.C., and Christie G.M., (1999) "Production line for planar SOFC ceramics: From laboratory to pre-pilot scale manufacturing," Fuel Cells Bulletin No. 14, November, Elsevier Science Ltd., ISSN 1464-2859, pp. 5-7.

Ishihara T., Matsuda, H., and Takita Y. (1994) "Doped LaGaO$_3$ perovskite type oxide as a new oxide ionic conductor", *Journal of the American Chemical Society*, vol. 116, No. 9, pp 3801-3803

Jalan V., Poirier J., Desai M., Morrisean B., (1990) "Development of CO and H2S tolerant PAFC anode catalysts", Proceedings of the second annual fuel cell contractors review meeting.

Khankar A., Elangovan S, Hartvigsen J., Rowley D., and Tharp M., (1998) "Recent progress in SOFCo's planar SOFC development," *Proceedings of the Fuel Cell Seminar*, November 16-19, Palm Springs CA, US.

Kinoshita K., (1988) *Carbon: Electrochemical and Physicochemical Properties*, Wiley Interscience, New York, NY.

Kunz H.R. (1987), "Transport of electrolyte in molten carbonate fuel cells" *Journal of the Electrochemical Society*, vol. 134, No. 1, pp.105 - 113.

Kordesch K.V., (1979) "Survey of Carbon and Its Role in Phosphoric Acid Fuel Cells," BNL 51418, prepared for Brookhaven National Laboratory.

Kortbeck, P.J., and Ottervanger, R., (1998) "Developing advanced DIR-MCFC co-generation for clean and competitive power" Abstracts of the Fuel Cell Seminar, Nov 16-19, 1998, Palm Springs, California.

Maru H.C., Marianowski, L.G.,(1976) Extended Abstracts, Abstract #31, Fall Meeting of the Electrochemical Society, October 17-22, Las Vegas, NV, Pg. 82, 1976.

Maru H.C. , Pigeaud A., Chamberlin R., Wilemski G., (1986) in Proceedings of the Symposium on Electrochemical Modeling of Battery, Fuel Cell, and Photoenergy Conversion Systems, edited by Selman J.R. and Maru H.C., The Electrochemical Society, Inc., Pennington, NJ,Pg. 398.

Maru H.C., Paetsch L., Pigeaud A. (1984) in Proceedings of the Symposium on Molten Carbonate Fuel Cell Technology, edited by R.J. Selman and T.D. Claar, The Electrochemical Society, Inc., Pennington, NJ, Pg. 20.

Mescal, C.M., (1999) "SOFC-PEM Hybrid Fuel Cell Systems," ETSU Report No. F/03/00177/REP, AEA Technology, Harwell UK.

Minh H., Anumakonda A., Chung B., Doshi R., Ferall J., Lear G., Montgomery K., Ong E., Schipper L., and Yamanis J', (1998) "High performance, reduced-temperature solid oxide fuel cell technology." Proceedings of the Fuel Cell Seminar, November 16-19, Palm Springs CA, US.

Mitteldorf J. and , Wilemski G. (1984) "Film thickness and distribution of electrolyte in porous fuel cell components", *Journal of the Electrochemical Society*, vol. 131, No. 8, pp 1784 - 1788

Petri R.J. and Benjamin T.G., (1986) in Proceedings of the 21st Intersociety Energy Conversion Engineering Conference, Volume 2, American Chemical Society, Washington, DC, Pg. 1156.

Pointon, K.D., (1997) "Review of work on internal reforming in the solid oxide fuel cell," ETSUreport F/01/00121/REP, AEA Technology, Harwell UK.

Sahibzada M., Rudkin R.A., Steele B.C.H., Metcalfe, I., and Kilner, J.A., (1997) "Evaluation of PEN structures incorporating supported thick film Ce 0.9 Gd 0.1 O1.95 electrolytes" Proceedings of the fith international symposium on Solid Oxide Fuel Cells (SOFC-V) pp. 244-253

Singhal S.C., (1997) "Recent Progress in Tubular Solid Oxide Fuel Cell Technology," Proceedings of the Fifth International Symposium on Solid Oxide Fuel Cells (SOFC-V), The Electrochemical Society, Inc., Pennington, NJ, U.S.

Stephenson D., & Ritchey I., (1997) "Parametric study of fuel cell and gas turbine combined cycle performance", Proceedings of the 1997 International Gas Turbine and Aeroengine Congress and Expostition (ASME conf. Code 473410)

Steele B.C.H., (1994) "State-of-the-art SOFC Ceramic Materials" Proceedings of the first European Solid Oxide Fuel Cell Forum, 3-7 October, Lucerne, Switzerland, Vol.1. pp 375-397.

Veyo S.E. & Forbes C.A. (1998) "Demonstrations based on Westinghouse's prototype commercial AES design", Proceedings of the third European Solid Oxide Fuel Cell Forum, pp79-86.

Vollmar H.-E., Maiaer C.-U., Nolscher C., Merklein T., and Pippinger M., (1999) "Innovative concepts for the co-production of electricity and syngas with solid oxide fuel cells, " Abstracts from the 6th Grove Fuel Cell Symposium, London, September 13-16, Elsevier Science Ltd., to be published in Journal of Power Sources.

Yuh C., Farooque, Johnsen R. (1992) "Understanding of carbonate fuel cell resistances in MCFCs", Proceedings of the Fourth Annual Fuel Cells Contractors Review Meeting, U.S.DOE/METC, pp 53-57

7

Fuelling Fuel Cells

7.1 Introduction

The fuel cell types that have been described in detail so far employ hydrogen as the preferred fuel, on account of its high reactivity for the electrochemical anode reaction, and because the oxidation of hydrogen produces water, which is environmentally benign. Vehicles employing PEM fuel cells running on hydrogen may therefore be termed "zero-emission", because the only emission from the vehicle is water. Unfortunately hydrogen does not occur naturally as a gaseous fuel, and so for practical fuel cell systems it usually has to be generated from whatever fuel source is locally available. Table 7.1 overleaf gives basic chemical and physical data on hydrogen and some of these other fuels that might be considered for fuel cells.

The next section deals with the characteristics of the various primary fossil fuels that can be used for the generation of hydrogen for fuel cell systems, ranging from petroleum and natural gas to coal. Bio-fuels are also a possible source of hydrogen gas, and their possibilities are discussed in Section 7.3. Sections 7.4 to 7.6 explain how such primary fuels may be converted to hydrogen, using different chemical conversion technologies. For stationary fuel cell power plants there are good arguments for carrying out the chemical conversion of fuel as close to the fuel cell stack as possible. The principal reason for this is that heat, which is generated in the fuel cell stack, may be used for part of the fuel processing. Heat integration in such systems is therefore an important part of system design and this is dealt with in Section 7.5. For transportation applications, the generation of hydrogen onboard vehicles presents particular system design considerations, and these are dealt with in Section 7.6.

Although hydrogen will normally be generated as required by the fuel processors described in Sections 7.4 to 7.6, there are times when hydrogen is produced in large central plants, and stored and transported for fuel cell use. There is already something of an infrastructure for producing and supplying hydrogen, as it is widely used in the chemical industry, in petroleum refining and ammonia manufacture for example. There are those who see hydrogen being widely used in this way as an energy vector (a method of storing and transporting energy) in the future, when we rely less on fossil fuels, and more on renewable sources of energy. However, for certain applications of fuel cells this may be a suitable way of providing fuel even now. This is especially so in the case of small, portable, low power fuel cell systems. In these circumstances the special and difficult problems involved with transporting and storing hydrogen come to the fore. Among these are the problems of safety, and the various mechanical and chemical ways in which the hydrogen can be held. These issues are explored and explained in Section 7.7.

Table 7.1 Some properties of hydrogen and other fuels considered for fuel cell systems

	Hydrogen	Methane	Ammonia	Methanol	Ethanol	Gasoline [a]
	H_2	CH_4	NH_3	CH_3OH	C_2H_5OH	C_8H_{18}
Molecular weight	2.016	16.04	17.03	32.04	46.07	114.2
Freezing point (°C)	-259.2	-182.5	-77.7	-97.8	-117.3	-56.8
Boiling point (°C)	-252.77	-161.5	-33.4	64.7	78.5	125.7
Net enthalpy of combustion at 25°C (kJ/mol)	241.8	802.5	316.3	638.5	1275.9	5512.0
Heat of vaporisation (kJ/kg)	445.6	510	1,371	1,100	855	368.1
Liquid density (kg/l)	77	425	674	792	789	702
Specific heat at STP ($J mol^{-1} K^{-1}$)	28.8	34.1	36.4	76.6	112.4	188.9
Flammability limits in air (%)	4 – 77	4 – 16	15 – 28	6 – 50	3-19	1 – 6
Autoignition temperature in air, (°C)	571	632	651	470	365	220

a - Gasoline is a blend of hydrocarbons and varies with producer, application and season. N-octane is reasonably representative of properties except vapour pressure., which is intentionally raised by introducing light fractions.

7.2 Fossil Fuels

7.2.1   Petroleum

Petroleum is a mixture of gaseous, liquid and solid hydrocarbon-based chemical compounds that occur in sedimentary rock deposits around the world. Crude petroleum has little value, but when it is refined it provides high value liquid feeds, solvents, lubricants and other products. Petroleum derived fuels account for up to one half of the world's total energy supply, and include gasoline, diesel fuel, aviation fuel, and kerosene. Simple distillation is able to separate various components of crude petroleum into generic fractions of different boiling ranges, as shown in Table 7.2.   The proportions of the different fractions that are obtained from any particular crude oil depend on the origin of supply.

Each fraction of petroleum contains a different proportion of chemical compounds, these being normal and branched paraffins or alkanes, monocyclic and polycyclic paraffins (naphthenes), mono-nuclear and polynuclear aromatic hydrocarbons.   Light naphthas contain principally normal alkanes and some monocyclic alkanes (e.g. cyclopentane and cyclohexane). As we move through from low boiling to higher boiling fractions so the proportion of these low molecular weight alkanes falls and the proportion of polycyclic alkanes and aromatic hydrocarbons increases. The various fractions can be used without further refining, but chemical modification is made to some fractions by the oil companies.

Hydro-refining, or hydrogenation, for example, is routinely carried out by refiners to reduce the aromatic content of some fractions. Gasoline is also doped with various compounds to improve engine lubrication, reduce corrosion, and reduce the risk of "knock" or pre-ignition in the internal combustion engine. It is also common to blend petroleum fractions and to carry out hydro-refining to ensure sufficiently high "octane" rating for different types of engine. In recent years oxygenates have been added to petroleum for the automobile to improve combustion characteristics in engines. More details of such petrochemical engineering is to be found in other textbooks on the subject of fuel technology. For fuel cells, we need to know something of the physical and combustion characteristics of the fuel, but it is also very important to understand the chemical composition of the fuel. It is the chemical composition that largely determines the type of fuel processing which may be used for generating hydrogen, as will become apparent in the following sections. It will be seen later that different technologies have to be used to convert the various fraction types into hydrogen for fuel cell systems. Especially of importance when the fuel is converted catalytically are the various trace compounds that may be present, since they may act as poisons for the conversion catalysts, and indeed also for the fuel cell stack. The most significant trace compounds in fossil fuels are mainly the organic compounds containing sulphur, nitrogen or oxygen and organo-metallic compounds, which include various porphyrins. The removal of contaminant sulphur compounds from raw fuels is dealt with in Section 7.4.

Table 7.2 Crude petroleum is a mixture of compounds that can be separated into different generic boiling fractions

Petroleum fraction	Boiling range (°C)
Light naphtha	-1 – 150
Gasoline	-1 – 180
Heavy naphtha	150 – 205
Kerosene	205 – 260
Stove oil	205 – 290
Light gas oil	260 – 315
Heavy gas oil	315 – 425
Lubricating oil	>400
Vacuum gas oil	425 – 625
Residue	>600

The gasoline fraction of petroleum, together with the heavier diesel, is widely distributed as a fuel for vehicles - ranging from passenger cars to heavy duty trucks and buses. Because the infrastructure for delivering such fuels is mature, there are good arguments for fuelling fuel cell vehicles using a similar fuel. However, "well to wheel" techno-economic studies indicate that the benefits of using gasoline in fuel cell vehicles compared with internal combustion engines are marginal compared to the high benefits of using hydrogen or even methanol. If such fuel is used for future fuel cell vehicles, it is will be of simpler composition than present vehicle fuels, since there are no reasons for adding components for anti-knock or lubrication purposes. It is also expected that in the future petroleum fuels will have a much lower sulphur content than those currently distributed.

7.2.2 Petroleum in mixtures: tar sands, oil shales, gas hydrates and LPG

There are various deposits of naturally occurring petroleum substances that can be found as a solid or near-solid material in sandstone at depths that are usually less than 2000 metres. Some may be found as an outcrop on the surface. Huge amounts of such tar sands can be found in various parts of the USA and Canada, but the high bitumen content (very high molecular weight) make recovery not as attractive as more conventional petroleum deposits. Similarly oil shales comprise a significant and largely untapped source of petroleum materials. Oil shales are compact laminated rocks of sedimentary origin in which organic petroleum is locked. The oil can be obtained from the rock by distillation. Of the world-wide shale oil deposits estimated to be over 20×10^{12} barrels only a small fraction $(3 \times 10^{12}$ barrels) is easily recoverable by conventional technologies. Much of the oil locked inside oil shales is of high molecular weight, and bituminous in nature. A discussion of the processing of such materials is outside the scope of this book.

In natural petroleum reservoirs where the prevailing pressures are high and temperatures low (under permafrost for example), methane and other normally gaseous hydrocarbons form ice-like hydrogen-bonded complexes (clathrates) with water. Such complexes are known as gas hydrates. Methane hydrates have also been considered as a method of transporting natural gas from remote fields. Other low molecular weight hydrocarbons, such as propane and butane, are often found associated in crude petroleum. When the crude petroleum material is distilled, these emerge as a gaseous product which is marketed as Liquefied Petroleum Gas or LPG. It is obtained as a refinery by-product, and is used extensively for applications as diverse as camping gas stoves and as a vehicle fuel. LPG is also attractive for fuel cell systems that may be used for remote stationary power sources where there is no pipeline source of gas, and possibly also for some vehicle applications.

7.2.3 Coal and coal gases

Coal is the most abundant of all fossil fuels, and chemically it is the most complex, being formed from the compaction and induration of various plant remains similar to those in peat. Coal is classified according to the kinds of inherent plant material (coal type), the degree of metamorphism (coal rank), and the degree of impurities (coal grade). It is worth pointing out that apart from combustion, further processing of coal to produce liquids, gases and coke is highly dependent on the properties of the raw coal material. For example, primary coking coals are those which have 20-30% volatile organic matter. Bituminous coal are best suited for carbonisation (heating in air to temperatures of 750-1500°C to form "coal gas" or "town gas"), whereas there are several processes for gasification of a variety of coal types.

Coal carbonisation was the original method for producing "town gas" in the nineteenth century. Simple carbonisation or distillation of coal yielded gas (a mixture of mainly hydrogen and carbon oxides), organic liquids (tars and phenolics) and a residual coke. Carried out in retorts of various types, carbonisation has been superseded for large-scale production by various coal gasification processes. In gasification, coal is usually reacted with steam and oxygen (or air) at high temperatures. The products of primary coal gasification are mainly gases, together with smaller amounts of liquids and solids. The relative proportion of products depends on the type of coal, the temperature and pressure of reaction and the relative amounts of steam or oxygen injected into the gasifier. Further processing of the raw gasifier product gas can be carried out, for example, to increase the methane content or alter the hydrogen/carbon monoxide ratio depending on what syngas is required.

The numerous coal gasification systems available today can be classified as one of three basic types: 1) moving-bed, 2) fluidised-bed, and 3) entrained-bed. All three types use steam, and either air or oxygen to partially oxidise coal into a gaseous product. The moving-bed gasifiers produce a low temperature (450 to 650 °C) gas containing methane and ethane arising from devolatilisation of the coal, together with a hydrocarbon liquid stream containing naphtha, tars, oils, and phenolic liquids. Entrained-bed gasifiers produce gas at high temperature (>1200 °C), virtually no devolatilisation products in the product gas stream, and much lower amounts of liquid hydrocarbons. In fact, the entrained-bed gas product is composed almost entirely of hydrogen, carbon monoxide and carbon dioxide. The product gas from the fluidised-bed gasifier falls somewhere between these two other reactor types in composition and temperature (925 to 1050 °C). In all of these gasifiers, the heat required for gasification (the reaction of coal and steam) is effectively provided by the partial oxidation of the coal. The temperature, and therefore composition, of the product gas is dependent upon the amount of oxidant and steam, as well as the design of the reactor that each process utilises. Gasifier product gases invariably contain contaminants that need to be removed before they can be used for fuel cells. Contaminant removal is therefore essential, and methods of gas clean up are described in the next section. Table 7.3 shows the compositions of some raw coal gases produced by some of the leading types of coal gasifier. Fuel cell developers have for many years recognised the benefits of fuelling fuel cells with the gases produced from such coal gasifiers, an example study being that of Brown et al. (1996).

Table 7.3 Typical coal gas compositions (mole percent basis)

	BG-Lurgi gasifier (non-slagging)	BG-Lurgi slagging gasifier	Moving bed O_2-blown (Lurgi)	Fluidised bed (Winkler)	Entrained bed O_2 blown (Texaco)	Entrained bed air-blown	Entrained bed O_2 blown (Shell)
Coal	Pittsburg 8	Pittsburg 8	Illinois no.6	Texas Lignite	Illinois no.6	Illinois no.6	Illinois no.6
Ar	trace	Trace	trace	0.7	0.9	trace	1.1
CH_4	8.5	7.2	3.3	4.6	0.1	1.0	
C_2H_6	0.7	0.1	0.1				
C_2H_4	0.3	0.2	0.2				
H_2	29.1	39.0	21.0	28.3	30.3	9.0	26.7
CO	18.0	55.5	5.8	33.1	39.6	16	63.1
CO_2	31.1	3.9	11.8	15.5	10.8	6	1.5
N_2	2.4	4.0	0.2	0.6	0.7	62	4.1
NH_4			0.4	0.1	0.1		
H_2O			61.8	16.8	16.5	5.0	2.0
H_2S			0.5	0.2	1.0		1.3
Total	100.0	100.0	100.0	100.0	100.0	100.0	100.0

7.2.4   Natural gas

Natural gas is the combustible gas that occurs in porous rocks in the earth's crust. It is found with or close to crude oil reserves, but may also occur alone in separate reservoirs. Most commonly, it forms a gas cap trapped between liquid petroleum and an impervious rock layer (cap rock) in a petroleum reservoir. If the pressure is high enough, the gas will be intimately mixed with or dissolved in the crude petroleum.

Chemically natural gas comprises a mixture of hydrocarbons of low boiling point. Methane is the component usually present in the greatest concentration, with smaller amounts of ethane, propane etc. In addition to hydrocarbons, natural gas contains various quantities of nitrogen, carbon dioxide and traces of other gases such as helium (often present in commercially recoverable quantities). Sulphur is also present to a greater or lesser extent, mostly in the form of hydrogen sulphide. The overall composition varies according to the source of the natural gas, and there are also seasonal variations. Natural gas is often described as being dry or lean (containing mostly methane), wet (containing considerable concentrations of higher molecular weight hydrocarbons), sour gas (with significant levels of H_2S), sweet gas (low in H_2S), and casinghead gas (derived from an oil well by extraction at the surface). Table 7.4 shows some typical compositions of natural gases from different regions around the world.

Table 7.4 Typical compositions of natural gases from different geographic regions

Component	North Sea	Qatar	Netherlands	Pakistan	Ekofisk	Indonesia
CH_4	94.86	76.6	81.4	93.48	85.5	84.88
C_2H_6	3.90	12.59	2.9	0.24	8.36	7.54
C_3H_8		2.38	0.4	0.24	2.85	1.60
i-C_4H_{10}	0.15	0.11		0.04	0.86	0.03
n-C_4H_{10}		0.21	0.1	0.06		0.12
C_{5+}		0.02		0.41	0.22	1.82
N_2	0.79	0.24	14.2	4.02	0.43	4.0
S	4 ppm	1.02	1 ppm	N/A	30 ppm	2 ppm

Values are % by volume unless otherwise stated

Some processing of the natural gas may be carried out close to the point of extraction before it enters a transmission system. Examples are the bulk removal of sulphur (sweetening), removal of high molecular weight hydrocarbons, nitrogen, acid gases, liquid water and liquid hydrocarbons. Nevertheless, there are wide variations in the composition of natural gas fed to transmission systems around the world. UK natural gas, for example, varies in composition according to the field from which it is supplied. Since natural gas composition does vary so much in composition, even with the season, it is common to enrich natural gas in some geographic regions by blending in mixtures of ethane, propane and butane for example. The requirements for this are usually dictated by the agreements between the gas supplier and the gas distributor. In addition, to meet the need for peak demand, enrichment with mixtures of propane/air or butane/air is also practised. The latter has important consequences for gas processing in fuel cell systems. For example the proprietary CRG catalyst, widely used around the world for steam reforming of natural gas, will only tolerate a small percentage of oxygen in the feed gas.

Natural gas has no distinctive odour (except for very sour gases), and for safety reasons, pipeline companies and utilities commonly odorise the gas either as the gas enters the transmission system, or within local distribution zones. Various odorants may be used, the most common being mixtures of thiophenes, and mercaptans. Tetrahydrothiophene (THT) is widely used throughout Europe and in the US (as Pennwall's odorant), whereas in the UK a cocktail of compounds is used comprising ethyl mercaptan, tertiary butyl mercaptan and di-ethyl sulphide.

7.3   Bio-fuels

Biomatter or biomass is a catch-all term for natural organic material associated with living organisms, including terrestrial and marine vegetable matter – everything from algae to trees, together with animal tissue and manure. On a global basis it is estimated that over 150 gigatonnes of vegetable biomatter are generated annually. The term 'biomass' is more often associated with production, expressed in tonnes per hectare, often also expressed in tonnes per hectare annual yield (t/ha). This yield ranges from about 13 t/ha for water hyacinth to 120 t/ha for napier grass. In view of its high energy content, biomass represents an important source of renewable fuel. This can be obtained by one of several routes:

- direct combustion
- conversion to biogas via pyrolysis, hydrogasifcation, or anaerobic digestion
- conversion to ethanol via fermentation
- conversion to syngas thermochemically, and then to methanol or ammonia
- conversion to liquid hydrocarbons by hydrogenation, or via the syngas/Fischer-Tropsch route

Another source of bio-fuel is from municipal waste. Again, apart from direct pyrolysis or incineration, gaseous fuels arising from landfill sites and other forms of refuse digestion can form a useful source of renewable energy, ideally suited to fuel cell systems. In the UK, approximately 80 percent of household waste is tipped, 10 percent is incinerated and 5 percent composted. The combustible component of such waste may be extracted chemically by conversion to gases, liquid distillates and char by anaerobic pyrolysis or biochemically by anaerobic digestion using methane-forming bacteria. Anaerobic digestion as currently practised requires a wet waste of relatively high nitrogen content, and the nitrogen/carbon ratio of about 0.03 is increased to 0.07 by the addition of animal manure, sewage sludge or other nitrogen-rich waste. Anaerobic digestors can be made at relatively small scale (a few kW), compared with pyrolysis gasifiers which normally only become attractive at the MW scale.

Bio-gases produced from biomass, land-fill, or anaerobic digestion contain mixtures of methane, carbon dioxide and nitrogen, together with various other organic materials. The compositions vary widely and, for example in the case of landfill, depend on the age of the site. A new site tends to produce gas with a high heating value, and this tends to decrease over a period of time. Some compositions of bio-gases are given in Table 7.5.

There is a particular attraction for using bio-gases in fuel cell systems. Most bio-gases have low heating values, caused by the relatively high levels of carbon oxides and nitrogen, making them unattractive for use in gas engines. However this drawback is not an issue for fuel cells particularly the MCFC and SOFC, which are able to handle very high concentrations of carbon oxides. This is also the case, but to a lesser extent for the PAFC. Developers have been assessing the use of bio-gases in such fuel cells for some time (Warren et al., 1986, and Langnickel, 1999).

Bio-liquids such as methanol and ethanol are also attractive bio-fuels for some fuel cell systems. Methanol is proposed as a fuel for fuel cell vehicles. Methanol can be synthesised from syngas that may be derived from biomass or natural gas. Ethanol can be produced directly by fermentation of biomass. Alcohols are attractive also because of the ease by which they can be reformed into hydrogen-rich gas. This is necessary if the reforming is to be carried out onboard a fuel cell vehicle.

Table 7.5 Example compositions of bio-gases

	Bio-gas[a]	Bio-gas[b]	Bio-gas[c]	Bio-gas[d]	Landfill gas[e]
Source	Agricultural sludge		Agricultural sludge	Brewery effluent	
Methane (vol %)	55-65	55-70	50-70	65-75	57
Ethane (vol %)		0			
Propane (vol %)		0			
Carbon dioxide (vol %)	33-43	30-45	30-40	25-35	37
Nitrogen (vol %)	2-1	0-2	small		6
Hydrogen sulphide (ppm)	<2000	~500	small	<5000	
Ammonia (ppm)	<1000	~100		<1	
Hydrogen (vol %)			small		
Water, relative moisture	80				
Higher Heating Value (MJ/nm^3)		23.3	>20		
Density (kg/nm^3)		1.16			

a) Paper BP-12 20th World Gas conference 1997.

b) Combined Utilization of Biogas and Natural Gas , J. Jemsen, S Tafdrup, and Johannes Chrisensen, Paper BO-06, 20th World Gas Conference, 1997.

c) Renewable Energy World, March 1999, page 75.

d) Caddet renewable energy newsletter, July 1999 pages 14-16 (Biogas used in Toshiba 200 kW phosphoric acid fuel cell.)

e) Caddet renewable energy Technical Brochure No. 32 (1996)

7.4   The Basics of Fuel Processing

7.4.1   Fuel cell requirements

Fuel processing may be defined as the conversion of the raw primary fuel supplied to a fuel cell system, into the fuel gas required by the stack. Each type of fuel cell stack has some particular fuel requirements, summarised in Table 7.6. Essentially, the lower the operating temperature of the stack, the more stringent are the requirements and the greater the demand placed on fuel processing. For example, fuel fed to a PAFC needs to be hydrogen-rich and contain less than about 0.5% carbon monoxide. The fuel fed to a PEM fuel cell needs to be essentially carbon monoxide free, while both the MCFC and SOFC fuel cells are capable of utilising carbon monoxide through the water gas shift reaction that occurs within the cell. Additionally, SOFCs and internal reforming MCFCs can utilise methane within the fuel cell whereas PAFCs and PEMs cannot. It is not widely known that PEMFCs can utilise some hydrocarbons, such as propane, directly although the performance is poor.

Table 7.6 The fuel requirements for the principal types of fuel cell

Gas species	PEM Fuel Cell	AFC	PAFC	MCFC	SOFC
H_2	Fuel	Fuel	Fuel	Fuel	Fuel
CO	Poison (>10ppm)	Poison	Poison (>0.5%)	Fuel [a]	Fuel [a]
CH_4	Diluent	Diluent	Diluent	Diluent [b]	Diluent [b]
CO_2 and H_2O	Diluent	Poison [c]	Diluent	Diluent	Diluent
S (as H_2S and COS)	Few studies, to date	Unknown	Poison (>50 ppm)	Poison (>0.5 ppm)	Poison (>1.0 ppm)

a – In reality CO reacts with H_2O producing H_2 and CO_2 via the shift reaction (7.3) and CH_4 with H_2O reforms to H_2 and CO faster than reacting as a fuel at the electrode.

b – A fuel in the internal reforming MCFC and SOFC.

c – The fact that CO_2 is a poison for the alkaline fuel cell more or less rules out its use with reformed fuels

Considerable research has been carried out in the field of fuel processing and reviews of some of the key technologies are readily available (e.g. Dicks, 1996). The following sections are intended to provide a basic explanation of the various technologies. Some detailed design of individual reactors and systems are proprietary, of course, but there is a wealth of information also available from various organisations involved in the development of fuel cell systems. Much of what is discussed here may be regarded as the practical application of chemical engineering to fuel cell systems.

7.4.2   Desulphurisation

Natural gas and petroleum liquids contain organic sulphur compounds that normally have to be removed before any further fuel processing can be carried out. In the case of natural gas the sulphur compounds may only be the odorants that have been added by the utility company for safety reasons. With petroleum fractions, the compounds may be highly aromatic in nature. Careful design of a desulphurisation system is required to ensure that the fuel gas passing through to the reformer catalyst or fuel cell stacks contains only very low levels of sulphur (typically below 0.1 ppm). If the fuel cell plant has a source of hydrogen-rich gas (usually from the reformer exit) it is common practice to recycle a small amount of this back to a HydroDesulphurisation (HDS) reactor. In this reactor, any organic sulphur containing compounds are converted, over a supported nickel-molybdenum oxide or cobalt-molybdenum oxide catalyst, into hydrogen sulphide via hydrogenolysis reactions of the type:-

$$(C_2H_5)_2 S + 2H_2 \rightarrow 2C_2H_6 + H_2S \qquad\qquad [7.1]$$

The rate of hydrogenolysis increases with increasing temperature and under operating temperatures of 300-400 $^\circ$C and in the presence of excess hydrogen the reaction is normally complete. It is worth pointing out that the lighter sulphur compounds easily undergo hydrogenolysis, whilst thiophene (C_4H_4S) and tetra-hydro-thiophene (THT) ($C_4H_8O_2S$) react with more difficulty, with a much slower reaction rate. The H_2S that is formed by such reactions is subsequently absorbed onto a bed of zinc oxide, forming zinc sulphide:

$$H_2S + ZnO \rightarrow ZnS + H_2O \qquad\qquad [7.2]$$

HDS is practised widely in industry and commercial catalysts and absorbents are available. The operating conditions and feed gas composition determine the choice of between nickel or cobalt catalysts. For example, nickel is a more powerful hydrogenation catalyst than cobalt and can cause more hydrocracking. Such undesirable hydrocracking reactions can occur over HDS catalysts if they become reduced to the metals by the hydrogen-rich gas, and the concentrations of organic sulphur compounds are very low. Alkene hydrogenation can also cause problems if such compounds are present in significant quantities. Such problems may occur with some liquid petroleum feedstocks but are unlikely to happen with natural gases. The optimum temperature for most HDS catalysts is between 350°C and 400°C, and the catalyst and zinc oxide may be placed in the same vessel. A variation on the HDS process, known as the PURASPEC™ process, is marketed by Synetix, part of the ICI group.

HDS as a means of removing sulphur to very low levels is ideally suited to PEM or PAFC systems. Unfortunately, HDS cannot easily be applied to internal reforming MCFC or SOFC systems, since there is no hydrogen-rich stream to feed to the HDS reactor. Most developers of such systems have therefore opted for removing sulphur from the feed gases using an absorbent. Activated carbon is especially suitable for small systems and can be impregnated with metallic promoters to enhance the absorption of specific materials such as H_2S. Molecular sieves may also be used, and Okada et al., 1994) have developed a mixed oxide absorbent system for fuel cells. The absorption capacity of materials such as active carbon is, however, quite low, and the beds of absorbent need replacing at regular intervals. This may be a serious economic disadvantage for large systems.

7.4.3   Steam reforming

Steam reforming is a mature technology, practised industrially on a large scale for hydrogen production and several detailed reviews of the technology have been published (Van Hook, 1981, Rostrup-Nielsen, 1984 and 1993), and useful data for system design is provided by Twigg (1989). The basic reforming reactions for methane and a generic hydrocarbon C_nH_m are:

$$CH_4 + H_2O \rightarrow CO + 3H_2 \qquad [\Delta H = 206 \text{ kJ mol}^{-1}] \qquad [7.3]$$

$$C_n H_m + n H_2 O \rightarrow n CO + (\tfrac{m}{2} + n) H_2 \qquad\qquad [7.4]$$

$$CO + H_2 O \rightarrow CO_2 + H_2 \qquad\qquad [\Delta H = -41 \text{ kJ mol}^{-1}] \qquad [7.5]$$

The reforming reactions (7.3 and 7.4), more correctly termed oxygenolysis reactions, and the associated water-gas shift reaction (7.5) are carried out normally over a supported nickel catalyst at elevated temperatures, typically above 500°C. Reactions 7.3 and 7.5 are reversible and normally reach equilibrium over an active catalyst, as at such high temperatures the rates of reaction are very fast. Over a catalyst that is active for reaction 7.3, reaction 7.5 nearly always occurs as well. The combination of the two reactions taking place means that the overall product gas is a mixture of carbon monoxide, carbon dioxide and hydrogen, together with unconverted methane and steam. The actual composition of the product from the reformer is then governed by the temperature of the reactor (actually the outlet temperature), the operating pressure, the composition of the feed gas, and the proportion of steam fed to the reactor. Graphs and computer models using thermodynamic data are available to determine the composition of the equilibrium product gas for different operating conditions. Figure 7.1 below is an example, showing the composition of the output at 1 bar.

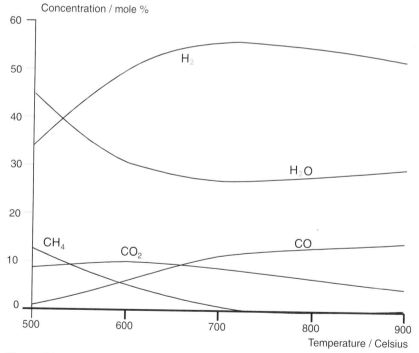

Figure 7.1 Equilibrium concentrations of steam reformation reactant gases as a function of temperature.

It can be seen that in the case of reaction 7.3, there are three molecules of carbon monoxide and one molecule of hydrogen produced for every molecule of methane reacted.

Le Chatelier's principle therefore tells us that the equilibrium will be moved to the right (i.e. in favour of hydrogen) if the pressure in the reactor is kept low. Increasing the pressure will favour formation of methane, since moving to the left of the equilibrium reduces the number of molecules. The effect of pressure on the equilibrium position of the shift reaction (7.5) is very small.

Another feature of reactions 7.3 and 7.4 is that they are usually very *endothermic* which means that heat needs to be supplied to the reaction to drive it forward to produce hydrogen and carbon monoxide. Higher temperatures (up to 700 °C) therefore favour hydrogen formation, as shown in Figure 7.1.

It is important to note at this stage that although the shift reaction (7.5) does occur at the same time as steam reforming, at the high temperatures needed for hydrogen generation, the equilibrium point for the reaction is well to the left of the equation. The result is that by no means all the carbon monoxide will be converted to carbon dioxide. For fuel cell systems that require low levels of CO, further processing will be required. This is explained in Section 7.4.9 below.

It is important to realise also that steam reforming is not always endothermic. For example, in the case of steam reforming a petroleum hydrocarbon such as naphtha, with the empirical formula $CH_{2.2}$, the reaction is most endothermic at the limit when the whole of the carbon is reformed to give oxides of carbon and hydrogen. This is the case when the reaction is carried out at relatively high temperatures. It is less endothermic and eventually exothermic (liberates heat) as the temperature is lowered. This is because as the temperature is lowered, so the reverse of reaction 7.1 becomes favoured, i.e. a competing reaction, namely the formation of methane, starts to dominate. This effect is illustrated in Table 7.7.

Table 7.7 Examples of typical heats of reaction in naphtha reforming at different temperatures, pressure and steam/carbon ratios. (Twigg, 1989)

Pressure kPa	Temp °C	Steam Ratio*	Reaction	ΔH (25°C) (kJ/mol $CH_{2.2}$)
2070	800	3.0	$CH_{2.2} + 3H_2O \rightarrow$ $0.2CH_4 + 0.4CO_2 + 0.4CO + 1.94H_2 + 1.81H_2O$	+102.5
2760	750	3.0	$CH_{2.2} + 3H_2O \rightarrow$ $0.35CH_4 + 0.4CO_2 + 0.25CO + 1.5H_2 + 1.95H_2O$	+75.0
31050	450	2.0	$CH_{2.2} + 2H_2O \rightarrow$ $0.75CH_4 + 0.25CO_2 + 0.14H_2 + 1.5H_2O$	- 48.0

* Steam/carbon ratio is the ratio between the number of moles of steam and the number of moles of carbon in the steam + fuel fed to the reformer reactor.

The implications of this for fuel cell systems are as follows. Natural gas reforming will invariably be endothermic, and heat will need to be supplied to the reformer at sufficiently high temperature to ensure a reasonable degree of conversion. Naphtha, and similar fractions (gasoline, diesel and logistics fuels) will also react under endothermic conditions if hydrogen is the preferred product and the operating temperature is kept high. However, if the reforming of naphtha is carried out at modest temperatures (up to 500-600°C), then the reactions will tend to yield greater concentrations of methane. More importantly the need for external heating will diminish as the reaction becomes exothermic. This may have important consequences for the reforming of gasoline as envisaged in some automotive fuel cell systems.

Another type of reforming is known as dry reforming, or CO_2 reforming, which can be carried out if there is no ready source of steam:-

$$CH_4 + CO_2 \rightarrow 2CO + 2H_2 \qquad [\Delta H = 247 \text{ kJ mol}^{-1}] \qquad [7.6]$$

This reaction may occur in internal reforming fuel cells, for example the MCFC, when anode exhaust gas containing carbon dioxide and water is recycled to the fuel cell inlet.

Hydrocarbons such as methane are not the only fuels suitable for steam reforming. Alcohols will also react in an oxyenolysis or steam reforming reaction, for example methanol:

$$CH_3OH + H_2O \rightarrow 3H_2 + CO_2 \qquad [\Delta H = 49.7 \text{ kJ mol}^{-1}] \qquad [7.7]$$

The mildly endothermic steam reforming of methanol is one of the reasons why methanol is finding favour with vehicle manufacturers as a possible fuel for fuel cell vehicles. Little heat needs to be supplied to sustain the reaction, which will readily occur at modest temperatures (e.g. 250°C) over catalysts of mild activity such as copper supported on zinc oxide. Notice also that carbon monoxide does not feature as a principal product of methanol reforming. This makes methanol reformate particularly suited to PEM fuel cells, where carbon monoxide, even at the ppm level, can cause substantial losses in performance due to poisoning of the platinum anode electrochemical catalyst. However, it is important to note that although carbon monoxide does not feature in reaction 7.7, this does not mean that it is not produced at all. The water gas shift reaction of 7.5 is reversible, and carbon monoxide in small quantities is produced. The result is that the carbon monoxide removal methods described in Section 7.4.9 below are still needed with PEM fuel cells, though the CO levels are low enough for PAFC.

7.4.4   Carbon formation and pre-reforming

One of the most critical issues in fuel cell systems is the risk of carbon formation from the fuel gas. This can occur in several areas of the system where hot fuel gas is present. Natural gas, for example, will decompose when heated in the absence of air or steam at temperatures above about 650°C via pyrolysis reactions of the type:

$$CH_4 \rightarrow C + 2H_2 \qquad [\Delta H = 75 \text{ kJ mol}^{-1}] \qquad [7.8]$$

Similar reactions can be written out for other hydrocarbons. Higher hydrocarbons tend to decompose more easily than methane and therefore the risk of carbon formation is higher with vaporised liquid petroleum fuels than with natural gas. Another source of carbon formation is from the disproportionation of carbon monoxide via the so-called Boudouard reaction:-

$$2CO \rightarrow C + CO_2 \qquad [7.9]$$

This reaction is catalysed by metals such as nickel and therefore there is a high risk of it occurring on steam reforming catalysts which contain nickel, and on nickel-containing stainless steel used for fabricating the reactors. Fortunately, there is a simple expedient to reduce the risk of carbon formation from reactions 7.8 and 7.9, and that is to add steam to the fuel stream. The principal effect of this is to promote the shift reaction (7.5), which has the effect of reducing the partial pressure of carbon monoxide in the fuel gas stream. Steam also leads to the carbon gasification reaction, which is also very fast:-

$$C + H_2O \rightarrow CO + H_2 \qquad [7.10]$$

The minimum amount of steam that needs to be added to a hydrocarbon fuel gas to avoid carbon deposition may be calculated. The principle here is that it is assumed that a given fuel gas/steam mixture reacts via 7.3, 7.4 and 7.5 to produce a gas which is at equilibrium with respect to reactions 7.3 and 7.5 at the particular temperature of operation. The partial pressures of carbon monoxide and carbon dioxide in this gas are then used to calculate an equilibrium constant for the Boudouard reaction (7.9). This calculated equilibrium constant is then compared with what would be expected from thermodynamic calculation at the temperature considered. If the calculated constant is greater than the theoretical one, then carbon deposition is predicted on thermodynamic grounds. If the calculated constant is lower than theory predicts, then the gas is said to be in a safe region and carbon deposition will not occur. In practice a steam/carbon ratio of 2.0 to 3.0 is normally employed in steam reforming systems so that carbon deposition may be avoided with a margin of safety.

A particular type of carbon formation occurs on metals, known as carburisation, leading to spalling of metal in a phenomenon known as "metal dusting". Again, it is important to reduce the risk of this in fuel cell systems, and some developers have used copper-coated stainless steel in their fuel gas pre-heaters to keep the risk to a minimum.

Carbon formation on steam reforming catalysts has been the subject of intense study over the years and is well understood. Carbon formed via the pyrolysis (7.8) and the Boudouard (7.9) reactions adopts different forms which can be identified under an electron microscope. The most damaging form of carbon is the filaments that "grow" attached to nickel crystallites within the catalyst. Such carbon formation can be very fast. If the steam flow to the reformer reactor is shut down for some reason, the consequences can be disastrous, carbon formation occurring within seconds, leading to permanent breakdown and fouling of the catalyst and plugging of the reactor. Control of the fuel processing is therefore an important aspect of fuel cell systems, especially if the steam is obtained by recycling product from the fuel cell stack. Commercial steam reforming catalysts contain elements such as potassium, and molybdenum to reduce the risk of carbon formation, which would otherwise rapidly deactivate the catalyst.

Another procedure to reduce the risk of carbon formation in a fuel cell system is to carry out some pre-reforming of the fuel gas before it is fed to the reformer reactor. Pre-reforming is a term used commonly in industry to describe the conversion of high molecular weight hydrocarbons via the steam reforming reaction at relatively low temperatures (typically 250 to 500°C). This process step (also known as "sweetening " of the gas) is carried out before the main reforming reactions in a reactor which operates adiabatically, i.e. no heat is supplied to it or removed from it. The advantage of carrying out pre-reforming is that high molecular weight hydrocarbons, which are more reactive than methane, are converted into hydrogen preferentially. The gas from the exit of a pre-reformer therefore comprises mainly methane, with steam, together with small amounts of hydrogen and carbon oxides, depending on the temperature of the pre-reformer reactor.

The use of a pre-reformer in a fuel cell system is illustrated in Figure 7.2, which was designed for a SOFC demonstration. In this case the pre-reformer was built to remove not only the higher hydrocarbons from the natural gas, but also to convert some 15 percent of the methane into hydrogen. This amount of external/pre-reforming ensured that the anodes of the SOFC were kept in a reduced condition, and that not too much thermal stress was placed on the SOFC stack by only allowing 85% reforming of the methane inside the stack.

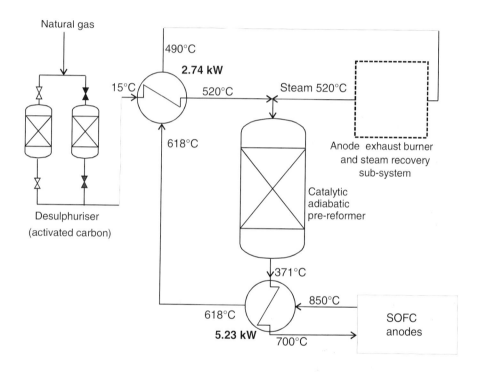

Figure 7.2 Pre-reformer system devised for Siemens 50 kW SOFC demonstration.

7.4.5   *Internal reforming*

Fuel cell developers have for many years known that the heat required to sustain the endothermic reforming of low molecular weight hydrocarbons (e.g. natural gas) can be provided by the electrochemical reaction in the stack. This has led to various elegant internal reforming concepts that have been applied to the molten carbonate or solid oxide fuel cells, on account of their high operating temperatures.

In contrast to the steam reforming reactions (7.3 and 7.4), the fuel cell reactions are exothermic, mainly due to heat production in the cell caused by internal resistances. Under practical conditions, with a cell voltage of 0.78 V, this heat evolved amounts to 470 kJ per mole methane. The overall heat production is about twice the heat consumed by the steam reforming reaction in an internally reforming fuel cell. Hence the cooling required by the cell, which is usually achieved by flowing excess gas through the cathode in the case of external reforming systems, will be much smaller for internal reforming systems. This has a major benefit on the electrical efficiency of the overall system.

Developers of internal reforming fuel cells have generally adopted one of two approaches; and these are usually referred to as direct (DIR) and indirect (IIR) internal reforming. They are illustrated schematically in Figure 7.3. In some cases a combination of both approaches has been taken. A thermodynamic analysis and comparison of the two approaches to internal reforming in the MCFC has recently been completed by Freni and Maggio (1997). Application of internal reforming offers several further advantages compared with external reforming:

- System cost is reduced because the separate external reformer is not needed.
- With DIR less steam is required (the anode reaction in the SOFC and MCFC produces steam).
- There is a more even distribution of hydrogen in a DIR cell that may result in a more even temperature distribution.
- The methane conversion is high, especially in DIR systems where the cell consumes hydrogen as it is produced.
- The efficiency of the system is higher. This is mainly because internal reforming provides an elegant method of cooling the stack, reducing the need for excess cathode air. This in turn lowers the requirement for air compression and recirculation.

Indirect internal reforming (IIR)
Also known as integrated reforming, this approach involves conversion of methane by reformers positioned in close thermal contact with the stack. An example of this type of arrangement alternates plate reformers with small cell packages. The reformate from each plate is fed to neighbouring cells. IIR benefits from close thermal contact between stack and reformer but suffers from the fact that heat is transferred well only from cells adjacent to the reformers and steam for the reforming must be raised separately. A variation of this type of arrangement places the reforming catalyst in the gas distribution path of each cell. With IIR the reforming reaction and electrochemical reactions are separated.

Direct internal reforming (DIR)
In direct internal reforming, the reforming reactions are carried out within the anode compartment of the stack. This can be done by placing reforming catalyst within the fuel cell channels in the case of the MCFC, but for the SOFC the high temperature of operation and anode nickel content of most SOFCs mean that the reactions can be performed directly on the anode. The advantage DIR is that not only does it offer good heat transfer, but there is also chemical integration - product steam from the anode electrochemical reaction can be used for the reforming without the need for recycling spent fuel. In principle, the endothermic reaction can be used to help control the temperature of the stack but this effect is not enough to completely offset the heat produced by the electrochemical reaction

and management of the temperature gradients is an issue. Examples of the use of direct internal reforming are to be found in the Siemens-Westinghouse tubular SOFC and in various DIR-MCFC concepts. These were described in Chapter 6.

Finally it should be pointed out that internal reforming may be applied to several hydrocarbon fuels such as natural gas and vaporised liquids such as naphtha and kerosenes. Logistic fuels have been also been demonstrated in internal reforming MCFC stacks, and coal-gases are particularly attractive for internal reforming MCFC and SOFC stacks, since not only are carbon monoxide and hydrogen consumed directly as fuels, but residual methane (e.g. from the BG-Lurgi slagging gasifier – see Table 7.3) is internally reformed.

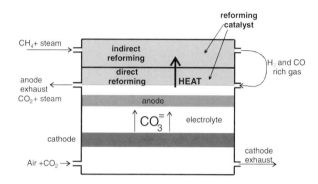

Figure 7.3 Schematic representation of direct and indirect internal reforming in the MCFC.

7.4.6   Direct hydrocarbon oxidation

A particular variation on internal reforming is that of direct hydrocarbon oxidation, also known as dry reforming. The Gibbs free energy change for the reaction of direct oxidation of methane to carbon dioxide and water is -796 kJ.mol^{-1} which is very close to the enthalpy change (ΔH = -802.5 kJ.mol^{-1}). In other words, if methane could be oxidised directly, most of the heat of reaction would be converted directly into electricity, with a maximum efficiency of :-

$$\frac{\Delta G}{\Delta H} = \frac{-796.5}{-802.5} \times 100 = 99.2\%$$

A great problem with direct hydrocarbon oxidation is that of carbon formation. As discussed in Section 7.4.4, this has a propensity to form on nickel-containing materials at temperatures even as low as 600°C. For this reason direct oxidation has been investigated on novel ceramic anodes for SOFCs made of mixed ionic- and electronic-conducting materials. A good example is the work recently reported by Perry Murray et al. (1999) on ceria-doped zirconia anodes in low temperature SOFCs. Other workers, e.g. Flot and Irvine (1999) are investigating various doped titanium oxide materials for similar duties.

Despite the attractiveness of direct hydrocarbon oxidation, development it is at an early stage. If steam is present with the fuel (and it will be generated, for example, on an SOFC anode), some steam reforming may occur. It would be easy to think that this would tend to offset the great advantage of direct methane oxidation, namely its high theoretical efficiency. In fact there is a substantial benefit in not having to supply steam to the fuel cell. Nevertheless, perhaps it is in the area of proton-conducting fuel cells that direct oxidation really does have a long term benefit. During developments of acid fuel cells in the 1960s several workers reported on the direct oxidation of hydrocarbons (Baker, 1965, Smith, 1966, and Liebhafsky and Cairns, 1968). Propane, for example, will decompose on platinum catalysts at moderate temperatures (below 200°C) to form protons, electrons, and CO_2 (via reaction with water from the aqueous acid electrolyte). As in the PEM fuel cell, the protons migrate to the cathode where they are oxidised with air and electrons to form water. It may be prudent for researchers now to revisit some of this earlier work to see if it holds the key for new ways of fuelling fuel cells.

7.4.7   Partial oxidation and autothermal reforming

As an alternative to steam reforming, methane and other hydrocarbons may be converted to hydrogen for fuel cells via partial oxidation:-

$$CH_4 + \tfrac{1}{2}O_2 \rightarrow CO + 2H_2 \qquad\qquad [\Delta H = - 247 \text{ kJ mol}^{-1}] \qquad\qquad [7.11]$$

Partial oxidation can be carried out at high temperatures (typically 1200 to 1500°C) without a catalyst. In this case it has the advantage over catalytic processes in that materials such as sulphur compounds do not need to be removed, although the sulphur does have to be removed at a later stage (as H_2S). High temperature partial oxidation can also handle much heavier petroleum fractions than catalytic processes, and is therefore attractive for processing diesels, logistic fuels and residual fractions. Such high temperature partial oxidation has been carried out at a large scale by several companies but it does not scale down well, and control of the reaction is problematic. If the temperature is reduced, and a catalyst employed then the process becomes known as Catalytic Partial Oxidation (CPO). Catalysts for CPO tend to be supported platinum-metal or nickel based. It should be noted that reaction 7.11 produces less hydrogen per molecule of methane than reaction 7.3. This means that partial oxidation (either non-catalytic or catalysed) is usually less efficient than steam reforming for fuel cell applications. Reaction 7.11 is effectively the summation of the steam reforming and oxidation reactions; about half of the fuel that is converted into hydrogen is oxidised to provide heat for the endothermic reforming reaction. Unlike the steam reforming reaction, no heat from the fuel cell can be utilised in the reaction, and the net effect is a reduced overall efficiency. Another disadvantage of partial oxidation occurs when air is used as the oxidant. This results in a lowering of the partial pressure of hydrogen at the fuel cell, because of the presence of the nitrogen. This in turn results in a lowering of the Nernst potential of the cell, again resulting in a lower system efficiency. To offset these negative aspects, a key advantage of partial oxidation is that it does not require steam. It may therefore be considered for applications where system simplicity is regarded as more important than high electrical conversion efficiency, for example small scale cogeneration, also known as micro-cogeneration.

Autothermal reforming is another commonly used term in fuel processing. This usually describes a process in which both steam and oxidant (oxygen, or more normally air) are fed with the fuel to a catalytic reactor. The advantages of autothermal reforming are that less steam is needed compared with conventional reforming and that all of the heat for the reforming reaction is provided by partial combustion of the fuel, so that no complex heat management engineering is required, resulting in a simple system design. As we shall see in Section 7.6, this is particularly attractive for mobile applications.

7.4.8   Hydrogen generation by pyrolysis or thermal cracking of hydrocarbons

An alternative to all of the above methods of generating hydrogen is to simply heat hydrocarbon fuels in the absence of air (pyrolysis). The hydrocarbon "cracks" or decomposes into hydrogen and solid carbon. The process is ideally suited to simple hydrocarbon fuels otherwise various by-products may be formed. The advantage of thermal cracking is that the hydrogen which is produced is very pure. The challenge is that the carbon that is also produced has to be removed from the reactor. This can be done by switching off the supply of fuel and admitting air to the reactor to burn off the carbon as carbon dioxide. The principle of switching the flow of fuel and oxidant is simple, but there are real difficulties, not least of which are the safety implications of admitting fuel and air into a reactor at high temperatures. Control of the pyrolysis is critical otherwise too much carbon will build up, and on a catalyst this can cause irreversible damage. If no catalyst is employed, again too much carbon may be formed. This may plug the reactor and mean that no flow of oxidising gas can be established to burn off the deposited material. Despite these substantial problems, pyrolysis is being considered seriously as an option for some fuel cell systems. Cracking of propane has been proposed recently to provide hydrogen for small PEM fuel cell systems (Ledjeff –Hey et al., 1999).

7.4.9   Further fuel processing – carbon monoxide removal

A steam reformer reactor running on natural gas and operating at atmospheric pressure with an outlet temperature of 800°C produces a gas comprising some 75% hydrogen, 15% carbon monoxide and 10% carbon monoxide on a dry basis. For the PEM fuel cell and the PAFC, reference to Table 7.6 shows that the carbon monoxide content must be reduced to much lower levels. Similarly, even the product from a methanol reformer operating at about 200 °C will have at least 0.1 % carbon monoxide content, depending on pressure and water content. This will be satisfactory for a PAFC, but is far too high for a PEMFC. The problem of reducing the carbon monoxide content of reformed gas streams is thus very important.

    We have seen that the water gas shift reaction:-

$$CO + H_2O \leftrightarrow CO_2 + H_2$$

takes place at the same time as the basic steam reforming reaction. However, the thermodynamics of the reaction are such that higher temperatures favour the production of carbon monoxide, and shifts the equilibrium to the left ($K = - 4.35$). The first approach is thus to *cool* the product gas from the steam reformer and pass it through a reactor containing catalyst, which promotes the shift reaction. This has the effect of converting carbon monoxide into carbon dioxide. Depending on the reformate composition more than one shift reactor may be needed, to reduce the carbon monoxide level to an acceptable level for the PAFC fuel cell. Iron-chromium catalyst is found to be effective for promoting the shift reactor at relatively high temperatures (400 - 500°C), and this may be followed by further cooling of the gas before passing to a second, low-temperature reactor (200 - 250°C) containing copper catalyst. At this temperature the proportion of carbon monoxide present will typically be about 0.25 to 0.5 %, and so these two stages of shift conversion are sufficient to decrease the carbon monoxide content to meet the needs of the PAFC. However, this level is equivalent to 2500 – 5000 ppm, which exceeds the limit for PEM fuel cells by a factor of about 100. It is similar to the CO content in the product from a methanol reformer.

For PEM fuel cells, further carbon monoxide removal is essential. This is usually done in one of three ways.

- In the **selective oxidation reactor** a small amount of air (typically around 2%) is added to the fuel stream, which then passes over a precious metal catalyst. This catalyst preferentially absorbs the carbon monoxide, rather than the hydrogen, where it reacts with the oxygen in the air. As well as the obvious problem of cost, these units need to be very carefully controlled. There is the presence of hydrogen, carbon monoxide and oxygen, at an elevated temperature, with a noble metal catalyst. Measures must be taken to ensure that an explosive mixture is not produced. This is a special problem in cases where the flowrate of the gas is highly variable, such as with a PEMFC on a vehicle.

- The **methanation** of the carbon monoxide is an approach that reduces the danger of producing explosive gas mixtures. The reaction is the opposite of the steam reformation reaction of equation 7.3, viz.:-

$$CO \; + \; 3H_2 \; \rightarrow \; CH_4 \; + \; H_2O \qquad\qquad (\Delta H = -206 \text{ kJ.mol}^{-1})$$

This method has the obvious disadvantage that hydrogen is being consumed, and so the efficiency is reduced. However, the quantities involved are small - we are reducing the carbon monoxide content from about 0.25 %. The methane does not poison the fuel cell, but simply acts as a diluent. Catalysts are available which will promote this reaction so that at about 200 °C the carbon monoxide levels will be less than 10 ppm. The catalysts will also ensure that any unconverted methanol is reacted to methane, hydrogen or carbon dioxide.

- **Palladium/platinum membranes** can be used to separate and purify the hydrogen. This is a mature technology that has been used for many years to produce hydrogen of exceptional purity. However, these devices are very expensive, and probably somewhat exceed the performance requirements.

These extra processes add considerably to the cost and complexity of the fuel processing systems for PAFC and PEMFC in comparison with those needed for MCFC and SOFC, a factor which we have already noted in Chapter 6.

7.5   Practical Fuel Processing – Stationary Applications

7.5.1   System designs for natural gas fed PEMFC and PAFC plants with steam reformers

In most stationary fuel cell plant, natural gas is the fuel of choice. It is widely available, clean and the fuel processing is generally straightforward. For the PEM fuel cell and the PAFC steam reforming is the preferred option, ensuring high overall fuel conversion efficiencies. In both cases the sulphur removal can be achieved by HDS.

The components of s suitable fuel processing system consist of an HDS unit for sulphur removal, steam reformer for conversion to hydrogen and carbon monoxide, and stages of high temperature and low temperature shift for conversion of carbon monoxide. For a PEMFC further carbon monoxide removal will be necessary. With such systems a degree of process integration is required, whereby heat from the fuel cell is utilised for various pre-heating duties. This is because the various chemical processes (desulphurisation, steam reforming, high temperature shift reaction, low temperature shift reaction) take place at different temperatures. There are thus a number of temperature changes to effect. The minimum required are:-

- Initial heating of the dry fuel gas to ~300 °C prior to desulphurisation
- Further heating prior to steam reforming at 600 °C or more
- Cooling of the reformer product gas to ~400 °C for the high temperature shift reaction
- Further cooling to ~200 °C for the low temperature shift reaction
- Possible heating or cooling prior to entry to fuel cell (depending on the fuel cell)
- Heating of water to produce the steam needed in the steam reformer

In addition to these six temperature changes, the steam reforming process requires heat at high temperature. This is usually provided by burning the anode exhaust gas from the fuel cell stack, which always contains some unconverted fuel. The reformer will corporate a burner in which this unused fuel is burnt. To obtain the high temperatures required by the reformer, it helps if the fuel gas from the fuel cell anode exhaust is further heated. Similarly, if the air needed by the burner is pre-heated it will be more effective. We can see that in such fuel cell systems, gases need to be both heated and cooled. Both processes can be combined without a net loss of heat by using heat exchangers. These were discussed briefly in Section 6.2.4 in the previous chapter.

A diagram of a fuel processing system that would be needed for a natural gas powered phosphoric acid fuel cell is shown in Figure 7.4. This needs a fuel gas at about 220 °C, with carbon monoxide levels down to about 0.5%. The system may look complicated enough, but even this has some simplifications. The following is an explanation of the process.

- The natural gas enters at around 20°C, and is heated in heat exchanger E, to a temperature suitable for desulphurization. Steam, sufficient for both reforming and the water gas shift reaction, is then added. The steam/methane mixture is then further heated before being passed to the steam reformer. Here it is heated to around 800/850 °C by the burner, and is converted to hydrogen and carbon monoxide, with some unreacted steam still present.

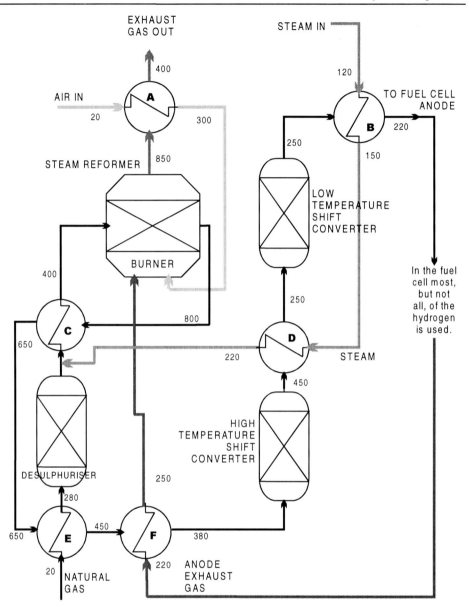

Figure 7.4 Diagram of a fuel processing system for a phosphoric acid fuel cell. The numbers indicate approximate likely temperatures.

- The reformate gas then passes through heat exchanger C again, losing heat to the incoming fuel gas. Further heat is lost to the incoming gas at E, and to the cathode exhaust gas at F.
- The gas is now sufficiently cool for the first shift converter, where the majority of the CO is converted to CO_2. At D the gas is further cooled, giving up heat to the incoming steam, before passing to the low temperature shift converter, where the final CO is

converted. The final cooling is accomplished at B, where the incoming steam is heated.

- The hydrogen rich fuel gas is then passed to the fuel cell. Here most, but not all, of the hydrogen is converted into electrical energy. The remaining gas returns at heat exchanger F, still at about 220°C. Here it is pre-heated prior to reaching the burner.

- Also coming to the burner is air, which will have been pre-heated by heat exchanger A, using energy from the burner exhaust gas.

- The steam arriving at about 120°C at heat exchanger B may have been created by heat from the fuel cell cooling system.

- The burner exhaust gas, still very hot, may be used to raise steam to drive any compressors needed to drive the process.

There are many other possible ways of configuring the gas flows and heat exchangers to get the required result, but they are unlikely to be simpler that this. As an example of another system, Figure 7.5 shows the schematic arrangement of the fuel processing within the Ballard 250 kW PEM fuel cell plant. Notice that we now have a selective oxidiser to further remove carbon monoxide. Another key difference is that this system operates at pressure, and the energy in the hot exhaust gases is used to drive the turbocompressors at the bottom of the diagram (see Chapter 8) rather than in heat exchangers.

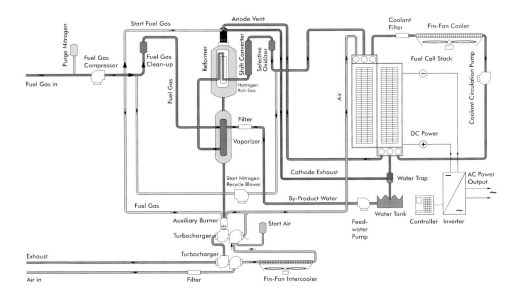

Figure 7.5 Process flow diagram for Ballard 250 kW PEMFC plant (reproduced by kind permission of Alstom Ballard GmbH)

7.5.2   Reformer and partial-oxidation designs

Compact regenerative reformers

In all of these systems the desulphuriser, shift reactors and carbon monoxide clean up systems are essentially packed-bed catalytic reactors of traditional design. However there are some reformers for stationary plant with novel features. Figure 7.6 shows an example of a reformer designed by Haldor Topsoe for PAFC systems in which the heat for the reforming reaction is provided by combustion of the lean anode exhaust gas supplemented if necessary by fresh fuel gas. In this reformer fuel is combusted at a pressure of some 4.5 bar in a central burner located in the bottom of a pressure vessel. Feed gas is fed downwards through the first catalyst bed where it is heated to around 675°C by convection from the combustion products and the reformed product gas both flowing counter current to the feed. On leaving the first bed of catalyst, the partially reformed gas is transferred through a set of tubes to the top of the second reforming section. The gas flows down through the catalyst, being heated typically to 830°C by convection from the co-currently flowing combustion products and also by radiation from the combustion tube. The combination of co-current and counter-current heat transfer minimises metal temperatures, an important consideration in high temperature reformer design. The advantages of such a reformer for fuel cell applications are: small size and suitability for small scale use; pressurised combustion of lean anode exhaust gas giving good process integration with the fuel cell; improved load following and lower cost.

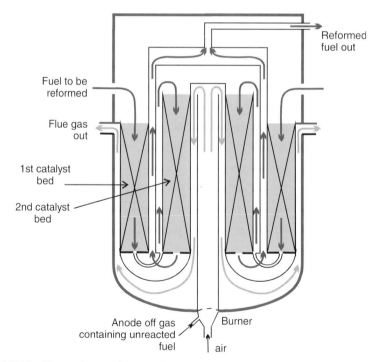

Figure 7.6 Haldor-Topsoe heat exchange reformer

Plate reformers

In the plate reformer a stack of alternate combustion and reforming chambers are separated by plates. The chambers are filled with suitable catalysts to promote the combustion and reforming reactions respectively. Alternatively, either side of each plate can be coated with combustion catalyst and reforming catalyst. The heat from the combustion reaction is used to drive the reforming reaction. Plate reformers have the advantage that they can be very compact and that they offer a means of reducing heat transfer resistance to a minimum. The use of a combustion catalyst means that low heating value gases (e.g. anode exhaust gases) can be burnt without the need for a supplementary fuel. The most advanced types of plate reformers use compact heat exchanger hardware in which catalyst is coated onto the exchanger surfaces or shims. Such devices are being developed by companies such as BG Technology Ltd in the UK and Pacific Northwest National Laboratory in the US.

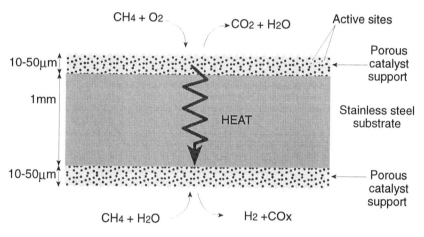

Figure 7.7 Plate reformer concept. Catalyst is coated as a thin film onto one or both sides of compact heat exchange material.

Membrane reactors

Hydrogen is able to permeate selectively through palladium or palladium alloy membranes. This has led to the demonstration in the laboratory of membrane reformers where hydrogen is selectively removed from the reformer as it is produced. This hydrogen removal increases the methane conversion for a given operating temperature above that which is predicted thermodynamically, and the hydrogen so produced is very pure, making it very suitable for PEM fuel cell systems. However, so far no commercial membrane reformer system has been produced.

Non-catalytic partial oxidation reactors

Non-catalytic partial oxidation is applied industrially by Texaco and Shell for the conversion of heavy oils to synthesis gas. In the case of the Shell basic partial oxidation process, liquid fuel is fed to a reactor together with oxygen, and steam. A partial combustion takes place in the reactor, which yields a product at around 1150°C. It is this high temperature which poses a particular problem for conventional partial oxidation. The reactor has to be made of expensive materials to withstand the high temperatures, and the product gas needs to be cooled to enable un-reacted carbon material to be separated from the gas stream. The high temperatures also mean that expensive materials of construction are required for the heat exchangers. In addition, the effluent from non-catalytic partial oxidation reactors invariably contain contaminants (including sulphur compounds) as well as carbon and ash which need to be dealt with. The complexity which these would add to a fuel cell system has meant that simple partial oxidation has not been a preferred option for fuel cell applications.

One interesting application of non-catalytic reactors has been the so-called plasma reformer. This type of reactor has the advantages of being compact, operating at moderate temperatures with fast start-up capability and good response to load changes. It is being developed in the research laboratory but has yet to find application in real fuel cell systems.

Catalytic partial oxidation (CPO) reactors

CPO reactors can be very simple in design, requiring only one bed of catalyst into which the fuel and oxidant (usually air) are injected. Often steam is added as well, in which case some conventional reforming also occurs. As mentioned before, the combination of CPO and steam reforming is often referred to as *autothermal reforming*, because there is no net heat supplied to or extracted from the reactor. All the heat for reforming is provided by partial combustion of the fuel. In such reactors two types of catalyst are sometimes used, one primarily for the CPO reaction, and the other to promote steam reforming, depending on the nature of the fuel and the application. Perhaps the most well-known CPO reactor is the Johnson Matthey HotSpot™ reactor (Edwards et al., 1998). A schematic representation of an early version of this technology is shown in Figure 7.8.

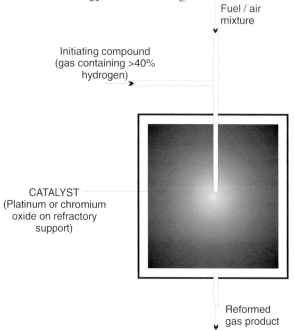

Figure 7.8 Simplified schematic of Johnson Matthey HotSpot™ reactor. Current versions of these devices have proprietary features not shown here.

It utilises a platinum/chromium oxide catalyst on a ceramic support. The novel feature of the reactor is the hot spot caused by point injection of the air-hydrocarbon mixture. This arrangement eliminates the need for pre-heating the fuel gas and air during operation, although for start-up on natural gas, the fuel should be pre-heated to around 500 °C. Alternatively, the reactor can be started from ambient temperature, by introducing an initiating fuel such as methanol or a hydrogen rich gas. Such initiating fuels are oxidised by air at ambient temperature over the catalysts, and this oxidation serves to raise the bed to the temperature needed for natural gas to react (typically over 450°C).

Several other developers have built CPO reactors for both mobile and stationary fuel cell applications. CPO of methanol proceeds with much lower heats of reaction compared with hydrocarbons and a very simple design of CPO reactor can be used. An example shown in Figure 7.9 is that developed by Kumar et al. (1996) at Argonne National Laboratory. In this reactor the catalyst is supported on a honeycomb monolith material similar to that found in automobile exhaust catalysts.

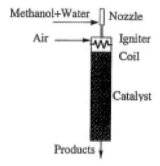

Figure 7.9 Argonne National Laboratory partial oxidation reformer

7.6   Practical Fuel Processing – Mobile Applications

7.6.1   General issues

There are two types of mobile applications for fuel cells. The first is where the fuel cell is providing the motive power for the vehicle, e.g. in combination with electric traction for cars and buses. For such Fuel Cell Vehicles (FCV) the PEM fuel cell (and for special applications the AFC) is the preferred option. The second application is where the fuel cell is providing only auxiliary power. An example of is the hotel load onboard ships, or the provision of supplementary electric power for trucks for heating and air-conditioning when stationary. Interestingly BMW has also announced a programme to develop SOFCs as an auxiliary power source for automobiles. However, it is when the fuel cell is providing the motive power, and fuel for the stack is processed onboard, that considerable extra demands are placed on the fuel processor.

In addition to the requirements for stationary power plant discussed earlier, onboard fuel processors for mobile applications need:-

*   to be compact (both in weight and volume)
*   to be capable of starting up quickly

- to be able to follow demand rapidly and operate efficiently over a wide operating range
- to be capable of delivering low-CO content gas to the PEM stack
- to emit very low levels of pollutants.

Over the past few years, research and development of fuel processing for mobile applications and small scale stationary applications has mushroomed. Many organisations are developing proprietary technology, but almost all of them are based on the options outlined in the previous section, namely steam reforming, CPO, or autothermal reforming. However, there is a vigorous debate being waged between vehicle manufacturers and fuel suppliers regarding the preferred fuel for FCVs. Methanol and gasoline are the front runners in the short to medium term, with hydrogen preferred by many as the long term option. Each fuel has its advantages and disadvantages, and it is likely that as FCVs do become commercial, the choice will be governed by both the type and duty of the vehicle, as well as the availability and distribution infrastructure for the fuel. Below we describe some of the processors that are being developed now for onboard processing of methanol and gasoline.

7.6.2   Methanol reforming

Leading developers of methanol reforming for vehicles at present are DaimlerChrysler, General Motors, Honda, International Fuel Cells, Mitsubishi, Nissan, Toyota, Johnson Matthey and Wellman CJB. Most are using steam reforming although Epyx and Wellman CJB, for example, are also working on partial oxidation. Two examples of practical methanol steam reforming systems are given below.

DaimlerChrysler methanol processor
Daimler-Benz (now DaimlerChrysler) developed a methanol processor for the NeCar 3 experimental vehicle. This was demonstrated in September 1997 as the world's first methanol-fuelled fuel cell car. It was used in conjunction with a Ballard 50 kW fuel cell stack. Characteristics of the methanol processor are given in Table 7.8, and Figure 7.10 shows how these were packaged in the car.

Table 7.8 Characteristics of the Methanol Processor for NeCar 3 (Kalhammer et al., 1998)

Maximum unit size	50 kWe
Power density	$1.1 kW_e.L^{-1}$ (reformer = 20 l, combustor = 5 l, CO selective oxidiser 20 l)
Specific power	$0.44 kW_e.kg^{-1}$ (reformer=34 kg, combustor=20 kg, CO sel. oxidiser=40 kg)
Energy efficiency	not determined
Methanol conversion	98-100%
Efficiency	
Turn-down ratio	20 to 1
Transient response	< 2 sec

Since the NeCar3 demonstration, Daimler Benz Ballard (DBB) have been working with BASF to develop a more advanced catalytic reformer system for vehicles. Already NeCar5 promises to house the fuel cell stack and reformer under the passenger compartment, as in Figure 7.11. DaimlerChrysler states that Necar 5 will have the full complement of back seats and the same, if limited, trunk space as the regular A-Class car on which the Necar series is based. DBB have also recently been working on a methanol-fuelled bus that uses modified NeCar 3 technology.

Figure 7.10 NeCar 3 packaging of stack and fuel processor (Reproduced by kind permission of DaimlerChrysler.)

General Motors/Opel methanol processor

General Motors European subsidiary Opel presented its first fuel cell powered vehicle, a concept car based on the Zafira, in September 1998. This uses a methanol reformer based on that developed by GM in a programme over many years working with Los Alamos National Laboratory in the US. The main features of the processor are given in Table 7.9

Table 7.9 Features of the GM Methanol Fuel Processor

Maximum unit size	30 kWe
Power density	$0.5 \text{kW}_e \text{L}^{-1}$
Specific power	$0.44 \text{kW}_e.\text{kg}^{-1}$
Energy efficiency	82-85%
Methanol conversion	>99%

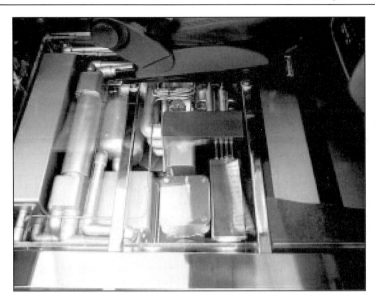

Figure 7.11 This methanol fuel processor fits below the passenger compartment, and a seat can just be seen at the top of the picture. Performance and technical details are proprietary (Reproduced by kind permission of DaimlerChrysler.)

Mercatox methanol processor

The EU-supported "Mercatox" project led by Wellman CJB in the UK has led to the development of a complete process for the on board steam reforming of methanol for vehicles. The basic premise is the use of compact, aluminium heat exchangers comprising corrugated plates coated on one side with methanol reforming catalyst and on the opposite side with combustion catalyst for burning, amongst other things, fuel cell off-gas. The principle is that of the plate reformer described in the previous section. A gas clean-up unit also uses the same basic construction with a selective oxidation catalyst coated on to the process side for converting carbon monoxide into carbon dioxide in the presence of hydrogen. The targets for a 50kWe system are a volume of 49 litres, mass of 50kg, a warm-up time of less than 5 seconds, transient response time with storage support of less than 5 seconds and without storage support of less than 50 milliseconds.

7.6.3 Methanol CPO/autothermal reforming

Several organisations are developing catalytic partial oxidation or autothermal reactors for methanol processing, examples being Hydrogen Burner Technology, Johnson Matthey (the HotSpot™ Fuel Processor), Argonne National Laboratory (US), Honda, DaimlerChrysler /Shell, and Epyx Corporation. The Johnson Matthey HotSpot™ fuel processor is shown below in Figure 7.12. Each one of the units is capable of producing in excess of 6000 litres of hydrogen per hour. It is left as an exercise for the reader to show, using equation A2.8 from Appendix 2, and the fact that the density of hydrogen is 0.084 kg.m^{-3} at NTP, to show that this is sufficient for a fuel cell of about 8 kW.

As mentioned before, the design for methanol CPO reactors can be very simple. There has been a push in recent years towards a multi-fuel processor, brought about by the requirements of the US Department of Energy to keep open all options for fuelling vehicles. Examples of multi-fuel processors are the integrated designs of North West Power Systems, LLC, (Edlund and Pledger, 1998).

Figure 7.12 The Johnson Matthey HotSpot™ reactor. These can be made in different forms suitable for methanol, methane or gasoline processing. (Picture by kind permission of Johnson Matthey Plc.)

7.6.4 Gasoline reforming

Companies such as Arthur D. Little, and later its spin-off Epyx Corporation in the US, have been developing reformers aimed at utilising gasoline type hydrocarbons (Teagan et al., 1998). It is felt that the adoption of gasoline as a fuel for FCVs is likely to find favour amongst oil companies, since the present distribution systems can be used. In the Epyx system the required heats of reaction for the reforming is provided by in-situ oxidising a fraction of the feedstock in a combustion (POX) zone. A nickel-based catalyst bed following the POX zone is the key to achieving full fuel conversion for high efficiency. The POX section operates at relatively high temperatures (1100-1500°C) whereas the catalytic reforming operates in the temperature range 800-1000°C. The separation of the POX and catalytic zones allows a relatively pure gas to enter the reformer, permitting the system to accommodate a variety of fuels. Shift reactors (high and low temperature) convert the product gas from the reformer so that the exit concentration of CO is less than 1%. As described earlier, an additional CO-removal stage is therefore needed to achieve the CO levels necessary for a PEM fuel cell. When designed for gasoline, the fuel processor also includes a compact desulfurisation bed integrated within the reactor vessel prior to the low temperature shift.

With all of the fuel processing options described for vehicular applications the assumption is generally made that onboard storage and processing of fuel will be needed. The main reason for this is to ensure a sufficient driving range for the vehicles, since we have come to expect this for our present day cars, lorries and buses. One way to side-step all of the problems associated with onboard fuel processing is to make the fuel processing plant stationary, and to store the hydrogen produced, which can be loaded onto the mobile system as required. In fact, this is may well be preferred option for some applications, such as buses, and leads us into the final section of this chapter.

7.7   Hydrogen Storage

7.7.1   Introduction to the problem

Up to now we have considered the production of hydrogen from fossil fuels on an "as needed" basis. The fuel itself has been the hydrogen store. However, there are times when it is more convenient and efficient to store the hydrogen fuel as hydrogen. This is particularly likely to be the case with low power applications, which would not justify the cost of fuel processing equipment. It can also be a reasonable way of storing electrical energy from sources such as wind driven generators and hydroelectric power, whose production might well be out of line with consumption. Electrolysers are used to convert to the electrical energy to hydrogen during times of high supply and low demand.

A small local store of hydrogen is also essential in the use of fuel cells for portable applications, unless the direct methanol fuel cell is being used.

As a result of its possible importance in the world energy scene as a general purpose energy vector, a great deal of attention has been given to the very difficult problem of hydrogen storage. The difficulties arise because although hydrogen has one of the highest specific energies (energy per kilogram) - which is why it is the fuel of choice for space missions - its density is very low, and it has one of the lowest energy densities (energy per cubic metre). This means that to get a large mass of hydrogen into a small space very high pressures have to be used. A further problem is that, unlike other gaseous energy carriers, it is very difficult to liquefy. It cannot be simply compressed, in the way that LPG or butane can. It has to be cooled down to about 22K, and even in liquid form its density is quite low, 71 $kg.m^{-3}$.

Although hydrogen can be stored as a compressed gas or a liquid, chemical methods can also be used. There are many compounds that can be manufactured that hold, for their mass, quite large quantities of hydrogen. To be useful these compounds must pass three tests:-

1. It must be possible to very easily make these compounds give up their hydrogen – otherwise there is no advantage over using a reformed fuel in one of the ways already described in this chapter.
2. The manufacturing process must be simple and use little energy – in other words the energy and financial costs of putting the hydrogen into the compound must be low.
3. They must be safe to handle.

A large number of chemicals that show promise have been suggested or tried. Some of these, together with their key properties, are listed in Table 7.10 below. Many of them do not warrant a great deal of consideration, as they fail one or more of the three tests above. Hydrazine is a good example. It passes the first test very well, and it has been used in fuel cells with great success, and was mentioned in connection with alkaline fuel cells in Section 5.2.3. However, hydrazine is both highly toxic and very energy intensive to manufacture, and so fails the second and third tests. Nevertheless, several of these compounds are used in practice, and will be described in more detail. The most important of these involve the use of metal hydrides. There are basically two forms. In one case the metal is a "rare earth" metal, and the hydride compound is reversible - in the right conditions it simply "gives up" the hydrogen. The other form involves the use of alkali metal hydrides; these are reacted with water, giving off hydrogen gas.

Table 7.10 Potential hydrogen storage materials. The "volume to store 1 kg" of H_2 figure excludes all the extra equipment needed to hold or process the compound, so it is not a practical figure. For example, all the alkali metal hydrides need large quantities of water, from which some of the hydrogen is also released. (See Section 7.7.6.)

Name	Formula	Percent hydrogen	Specific gravity	Vol. (L) to store 1 kg H_2	Notes
Simple hydrides					
Liquid H_2	H_2	100	0.07	14	Cold, -252°C
Lithium hydride	LiH	12.68	0.82	6.5	Caustic
Beryllium hydride	BeH_2	18.28	0.67	8.2	Very toxic
Diborane	B_2H_6	21.86	0.417	11	Toxic
Liquid methane	CH_4	25.13	0.415	9.6	Cold –175°C
Ammonia	NH_4	17.76	0.817	6.7	Toxic, 100 ppm
Water	H_2O	11.19	1.0	8.9	
Sodium hydride	NaH	4.3	0.92	25.9	Caustic, but cheap
Calcium hydride	CaH_2	5.0	1.9	11	
Aluminium hydride	AlH_3	10.8	1.3	7.1	
Silane	SiH_4	12.55	0.68	12	Toxic 0.1 ppm
Potassium hydride	KH	2.51	1.47	27.1	Caustic
Titanium hydride	TiH_2	4.40	3.9	5.8	
Complex hydrides					
Lithium borohydride	$LiBH_4$	18.51	0.666	8.1	Mild toxicity
Aluminium borohydride	$Al(BH_4)_3$	16.91	0.545	11	Mild toxicity
Lithium aluminium hydride	$LiAlH_4$	10.62	0.917	10	
Hydrazine	N_2H_4	12.58	1.011	7.8	Toxic 10 ppm
Hydrogen absorbers					
Palladium hydride	Pd_2H	0.471	10.78	20	
Titanium iron hydride	$TiFeH_2$	1.87	5.47	9.8	

Another method that is not yet practical, but might be feasible, is the use of carbon nanofibres. So far there have been no clear demonstrations of the feasibility of technology, so we will not describe it any further here – this might be one to watch in the future.

So, the principal methods of storing hydrogen that will be described in the following sections are:-

- Compression in gas cylinders
- Storage as a cryogenic liquid
- Storage as a reversible metal hydride
- The use of metal hydride reactions with water

None of these methods, which are all described in more detail later, is without considerable problems, and in each situation their advantages and disadvantages will play differently. However, before we do this we must address the vitally important issue of safety in connection with storing and using hydrogen.

7.7.2   Safety

Hydrogen is a unique gaseous element, possessing the lowest molecular weight of any gas. It has the highest thermal conductivity, velocity of sound, mean molecular velocity, and the lowest viscosity and density of all gases. Such properties lead hydrogen to have a leak rate through small orifices faster than all other gases. Hydrogen leaks 2.8 times faster than methane and 3.3 times faster than air. In addition hydrogen is a highly volatile and flammable gas, and in certain circumstances hydrogen and air mixtures can detonate. The implications for the design of fuel cell systems are obvious, and safety considerations must feature strongly.

Hydrogen therefore needs to be handled with care. Systems need to be designed with the lowest possible chance of any leaks, and should be monitored for such leaks regularly. However, it should be made clear that, all things considered, hydrogen is no more dangerous, and in some respects it is rather less dangerous than other commonly used fuels. Table 7.9 below gives the key properties relevant to safety of hydrogen and two other gaseous fuels widely used in homes, leisure and business - methane and propane.

From this table the major problem with hydrogen appears to be the minimum ignition energy, apparently indicating that a fire could be started very easily. However, all these energies are in fact very low, lower than those encountered in most practical cases. A spark can ignite any of these fuels. Furthermore, against this must be set the much higher minimum concentration needed for detonation - 18 % by volume. The lower concentration limit for ignition is much the same as for methane, and a considerably lower concentration of propane is needed. The ignition temperature for hydrogen is also noticeably higher than for the other two fuels.

Another potential hazard arises from the rather greater range of concentrations needed to cause detonation. This means that care must be taken to prevent the build-up of hydrogen in confined spaces. Fortunately, this is usually easily done; the high buoyancy, and high average molecular velocity ensures that hydrogen is the most rapidly dispersing of all gases.

A safety problem that can arise with hydrogen is that when it is burning the flame is virtually invisible. Fire-fighting teams will almost certainly have the necessary equipment to detect such fires, but this point should be borne in mind by non-specialists.

Considering these figures, hydrogen seems much the same as the other fuels from the point of view of potential danger. It is the much lower density that gives hydrogen a comparative advantage from a safety point of view. The density of methane is similar to air, which means it does not disperse quickly, but tends to mix with the air. Propane has a lower density than air, which tends to make it sink and collect at low points, such as in basements, in drains, and in the hulls of boats, where it can explode or set alight with devastating effects. Hydrogen on the other hand, is so light that it rapidly disperses upwards. This means that the concentration levels necessary for ignition or detonation are very unlikely to be achieved.

Table 7.11 Properties relevant to safety for hydrogen and two other commonly used gaseous fuels

	Hydrogen	Methane	Propane
Density, $kg.m^{-3}$ at NTP	0.084	0.65	2.01
Ignition limits in air, volume % at NTP	4.0 to 77	4.4 to 16.5	1.7 to 10.9
Ignition temperature, °C	560	540	487
Min. ignition energy in air, MJ	0.02	0.3	0.26
Max. combustion rate in air, $m.s^{-1}$	3.46	0.43	0.47
Detonation limits in air, volume %	18 to 59	6.3 to 14	1.1 to 1.3
Stoichiometric ratio in air	29.5	9.5	4.0

Figure 7.13 Icon of a myth. The "Hindenburg disaster" of 6th May 1937 put an end to the airship as a means of transport, and it has also been a major "public relations" problem for hydrogen, since this was the lifting gas used. The accident led to the widely held myth that hydrogen is a particularly dangerous substance. Although the accident was tragic for those involved, the number of casualties was 37, quite low for an aircraft crash. About 2/3 of those on board survived. Many of those who died were burnt by the diesel fuel for the propulsion system, and in any case the fire did not start with the hydrogen, but in the skin of the airship, which was made of a highly flammable compound. (Bain and VanVorst, 1999).

Hydrogen, like all fuels, must be carefully handled. However, taking all things into account, it does not present any greater potential for danger than any other flammable liquids or gases in common use today. In some applications, for example boats, it has many safety advantages over what is generally used at the moment. In the sections that follow we consider different storage methods. Some of these bring special safety hazards of their own, which we consider in turn.

7.7.3   The storage of hydrogen as a compressed gas

Storing hydrogen gas in pressurised cylinders in the most technically straightforward method, and the most widely used for small amounts of the gas. Hydrogen is stored in this way at thousands of industrial, research and teaching establishments, and in most locations local companies can readily supply such cylinders in a wide range of sizes. However, in these applications the hydrogen is nearly always a chemical reagent in some analytical or production process. When we consider using and storing hydrogen in this way as an energy vector, then the situation appears less satisfactory.

Two systems of pressurised storage are compared below. The first is a standard steel alloy cylinder at 200 bar, of the type commonly seen in laboratories. The second is for larger scale hydrogen storage on a bus, as described by Zieger (1994). This tank is constructed with a 6mm thick aluminium inner liner, around which is wrapped a composite of aramide fibre and epoxy resin. This material has a high ductility, which gives it good burst behaviour, in that is rips apart rather than disintegrating into many pieces. The burst pressure is 1200 bar, though the maximum pressure used is 300 bar.[1]

Table 7.12 Comparative data for two cylinders used to store hydrogen at high pressure. The first is a conventional steel cylinder, the second a larger composite tank for use on a hydrogen powered bus.

	2 L steel, 200 bar	147 L composite, 300 bar
Mass of empty cylinder	3.0 kg	100 kg
Mass of hydrogen stored	0.036 kg	3.1 kg
Storage efficiency (% mass H_2)	1.2 %	3.1 %
Specific energy	0.47 kWh.kg^{-1}	1.2 kWh.kg^{-1}
Volume of tank (approx.)	2.2 L (0.0022 m^3)	220 L (0.22 m^3)
Mass of H_2 per litre	0.016 kg.L^{-1}	0.014 kg.L^{-1}

The larger scale storage system is, as expected, a great deal more efficient. However, this is slightly misleading. These large tanks have to be held in the vehicle, and the weight needed to do this should be taken into account. In the bus described by Zieger (1994), which used hydrogen to drive an internal combustion engine, 13 of these tanks were mounted in the roof space. The total mass of the tanks and the bus structure reinforcements is 2550 kg, or 196 kg per tank. This brings down the "storage efficiency" of the system to 1.6%, not so very different from the steel cylinder. Another point is that in both systems we have ignored the weight of the connecting valves, and of any pressure reducing

---

[1] It should be noted that at present composite cylinders are about three times the cost of steel cylinders of the same capacity.

regulators. For the 2 L steel cylinder system this would typically add about 2.15 kg to the mass of the system, and reduce the storage efficiency to 0.7%. (Kahrom, 1999)

The reason for the low mass of hydrogen stored, even at such very high pressures, is of course its low density. The density of hydrogen gas at normal temperature and pressure is 0.084 kg.m^{-3}, compared to air, which is about 1.2 kg.m^{-3}. Usually less than 2% of the storage system mass is actually hydrogen itself.

Pressurised hydrogen gas is mainly used in fairly small quantities, and has to be transported with great care. For small scale users of fuel cells the figures quoted by Kahrom (1999), which take into account all costs, including depreciation of cylinders, administration, cost of pressure reducing valves and so on, estimate the cost of hydrogen as about $2.2 per g. Using the figures given in Appendix 2, Section A2.4, we can see that this corresponds to about $56 per kWh, or about $125 per kWh for the electricity from a fuel cell of efficiency 45%. This is absurdly expensive when compared to mains electricity, but is considerably cheaper than the cost of electricity from primary batteries

The metal that the pressure vessel is made from needs very careful selection. Hydrogen is a very small molecule, of high velocity, and so it is capable of diffusing into materials that are impermeable to other gases. This is compounded by the fact that a very small fraction of the hydrogen gas molecules may dissociate on the surface of the material. Diffusion of atomic hydrogen into the material may then occur which can affect the mechanical performance of materials in many ways. Gaseous hydrogen can build up in internal blisters in the material, which can lead to crack promotion (hydrogen induced cracking). In carbonaceous metals such as steel the hydrogen can react with carbon forming entrapped CH_4 bubbles. The gas pressure in the internal voids can generate an internal stress high enough to fissure, crack or blister the steel. The phenomenon is well known and is termed hydrogen embrittlement.. Certain chromium-rich steels and Cr-Mo alloys have been found that are resistant to hydrogen embrittlement. Composite reinforced plastic materials are also used for larger tanks, as has been outlined above.

As well as the problem of very high mass, there are considerable safety problems associated with storing hydrogen at high pressure. A leak from such a cylinder would generate very large forces as the gas is propelled out. It is possible for such cylinders to become essentially jet-propelled torpedoes, and to inflict considerable damage. Furthermore, vessel fracture would most likely be accompanied by autoignition of the released hydrogen and air mixture, with an ensuing fire lasting until the contents of the ruptured or accidentally opened vessel are consumed. (Hord, 1978) Nevertheless, this method is widely and safely used, provided the safety problems, especially those associated with the high pressure, are avoided by correctly following the due procedures. In vehicles, for example, pressure relief valves or rupture discs (see next section) are fitted which will safely vent gas in the event of a fire for example. Similarly, pressure regulators attached to hydrogen cylinders are fitted with flame-traps to prevent ignition of the hydrogen. The main advantages of storing hydrogen as a compressed gas are:-

- Simplicity
- Indefinite storage time
- No purity limits on the hydrogen

It is most widely used in places where the demand for hydrogen is variable, and not so high. It is also used for buses, both for fuel cells and internal combustion engines. It is well suited to storing the hydrogen from electrolysers that are run at times of excess electricity supply. One such system is shown in Figure 7.14.

Figure 7.14 Hydrogen generation and storage plant at Munich airport, Germany. The electrolyser is in the foreground, and the horizontal storage tanks can be just be seen behind on the right. (Photograph reproduced by kind permission of Hamburgische Electricitäts-Werke AG.)

7.7.4   Storage of hydrogen as a liquid

The storage of hydrogen as a liquid (commonly called LH_2), at about 22K, is currently the only widely used method of storing large quantities of hydrogen. A gas cooled to the liquid state in this way is known as a cryogenic liquid. Large quantities of cryogenic hydrogen are currently used in processes such as petroleum refining and ammonia production. Another notable user is NASA, which has huge 3200 m^3 (850,000 US gallon) tanks to ensure a continuous supply for the space programme. Figure 7.15 shows a tank for storing cryogenic hydrogen in Hamburg, Germany. Two fuel cell systems, of power about 200 kW, can also be seen to the right of the picture.

The hydrogen container is a large, strongly reinforced vacuum (or Dewar) flask. The liquid hydrogen will slowly evaporate, and the pressure in the container is usually maintained below 3 bar, though some larger tanks may use higher pressures. If the rate of evaporation exceeds the demand, then the tank is occasionally vented to make sure the pressure does not rise too high. A spring loaded valve will release, and close again when the pressure falls. The small amounts of hydrogen involved are usually released to the atmosphere, though in very large systems it may be vented out through a flare stack and burnt. As a back-up safety feature a rupture disc is usually also fitted. This consists of a ring covered with a membrane of controlled thickness, so that it will withstand a certain pressure. When a safety limit is reached, the membrane bursts, releasing the gas. However, the gas will continue to be released until the disc is replaced. This will not be done till all the gas is released, and the fault rectified.

When the LH_2 tank is being filled, and when fuel is being withdrawn, it is most important that air is not allowed into the system, otherwise an explosive mixture could form. The tank should be purged with nitrogen before filling.

Figure 7.15 LH_2 tank on the left. To the right of the picture are two approx. 200 kW fuel cell systems, providing heat and power to the houses and offices nearby. (Photograph reproduced by kind permission of Hamburgische Electricitäts-Werke AG)

Although usually used to store large quantities of hydrogen, considerable work has gone into the design and development of LH_2 tanks for cars, though this has not been directly connected with fuel cells. BMW, among other automobile companies, has invested heavily in hydrogen power internal combustion engines, and these have nearly all used LH_2 as the fuel. Such tanks have been through very thorough safety trials. The tank used in their hydrogen powered cars is cylindrical in shape, and is of the normal double wall, vacuum or Dewar flask type of construction. The walls are about 3 cm thick, and consist of 70 layers of aluminium foil interlaced with fibre-glass matting. The maximum operating pressure is 5 bar. The tank stores 120 litres of cryogenic hydrogen. The density of LH_2 is very low, about 71 kg.m^{-3}, so 120 litres is only 8.5 kg. (Reister and Strobl, 1992) The key figures are shown in the table below.

Table 7.13 Details of a cryogenic hydrogen container suitable for cars

Mass of empty container	51.5 kg
Mass of hydrogen stored	8.5 kg
Storage efficiency (% mass H_2)	14.2 %
Specific energy	5.57 kWh.kg^{-1}
Volume of tank (approx.)	0.2 m^3
Mass of H_2 per litre	0.0425 kg.L^{-1}

The hydrogen fuel feed systems used for car engines cannot normally be applied unaltered to fuel cells. One notable difference is that in LH_2 power engines the hydrogen is often fed to the engine still in the liquid state. If it is a gas, then being at a low temperature is an advantage, as it allows a greater mass of fuel/air mixture into the engine. For fuel cells, the hydrogen will obviously need to be a gas, and pre-heated as well. However, this is not a very difficult technical problem, as there is plenty of scope for using waste heat from the cell via heat exchangers.

One of the problems associated with cryogenic hydrogen is that the liquefaction process is very energy intensive. Several stages are involved. The gas is firstly compressed, and then cooled to about 78 K using liquid nitrogen. The high pressure is then used to further cool the hydrogen by expanding it through a turbine. (See Section 8.8.) An additional process is needed to convert the H_2 from the isomer where the nuclear spins of both atoms are parallel (ortho-hydrogen) to that where they are anti-parallel (para-hydrogen). This process is exothermic, and if allowed to take place naturally would cause boil-off of the liquid. In all, the energy required to liquefy the gas is about 40% of the specific enthalpy or heating value of the hydrogen.

Figure 7.16 Refilling with LH_2. Note that there are two pipes. The system is fully sealed to prevent inlet of air. One pipe carries in the liquid. The other draws off the hydrogen gas that is formed by the slowly evaporating liquid. (Photograph reproduced by kind permission of Hamburgische Electricitäts-Werke AG)

In addition to the regular safety problems with hydrogen, there are a number of specific difficulties concerned with cryogenic hydrogen. Frostbite is a hazard of concern. Human skin can easily become frozen or torn if it comes into contact with cryogenic surfaces. All pipes containing the fluid must be insulated, as must any parts in good thermal contact with these pipes. Insulation is also necessary to prevent the surrounding air from condensing on the pipes, as an explosion hazard can develop if liquid air drips onto nearby combustibles. Asphalt, for example, can ignite in the presence of liquid air. (Concrete paving is used around static installations.) Generally though, the hazards of hydrogen are somewhat less with LH_2 than with pressurised gas. One reason is that if there is a failure of the container, the fuel tends to remain in place, and vent to the atmosphere more slowly. Certainly, LH₂ tanks have been approved for use in cars in Europe. Figure 7.16 shows a van being filled with liquid hydrogen in a scheme for company vehicles operating in Hamburg, Germany. Although this vehicle uses an internal combustion engine, it shows that an infrastructure based on LH_2 is possible, and could be applied to fuel cells.

7.7.5   Reversible metal hydride hydrogen stores

Certain metals, particularly mixtures (alloys) of titanium, iron, manganese, nickel, chromium, and others, can react with hydrogen to form a metal hydride in a very easily controlled reversible reaction. The general equation is:-

$$M \; + \; H_2 \; \leftrightarrow \; MH_2 \qquad\qquad [7.12]$$

One example of such an alloy is the last entry is Table 7.10, titanium iron hydride. In terms of *mass* this is not a very promising material, it is the *volumetric* measure that is the advantage of these materials. It requires one of the lowest volumes to store 1 kg in Table 7.8, certainly it is one of the lowest practical materials. It actually holds more hydrogen per unit volume than pure liquid hydrogen[2].

To the right, the reaction of 7.12 is mildly exothermic. To release the hydrogen then, small amounts of heat must be supplied. However, metal alloys can be chosen for the hydrides so that the reaction can take place over a wide range of temperatures and pressures. In particular, it is possible to choose alloys suitable for operating at around atmospheric pressure, and room temperature.

The system then works as follows. Hydrogen is supplied at a little above atmospheric pressure to the metal alloy, inside a container. The reaction of 7.12 proceeds to the right, and the metal hydride is formed. This is mildly exothermic, and in large systems some cooling will need to be supplied, but normal air cooling is often sufficient. This stage will take a few minutes, depending on the size of the system, and if the container is cooled. It will take place at approximately constant pressure.

Once all the metal has reacted with the hydrogen, then the pressure will begin to rise. This is the sign to disconnect the hydrogen supply. The vessel, now containing the metal hydride, will then be sealed. Note that the hydrogen is only stored at modest pressure, typically up to 2 bar.

When the hydrogen is needed, the vessel is connected to, for example, the fuel cell. The reaction of 7.12 then proceeds to the left, and hydrogen is released. If the pressure rises above atmospheric, the reaction will slow down or stop. The reaction is now endothermic, so energy must be supplied. This is supplied by the surroundings – the vessel will cool slightly as the hydrogen is given off. It can be warmed slightly to increase the rate of supply, using, for example, warm water or the air from the fuel cell cooling system.

---

[2] This may at first sight seem impossible. However, contrary to what a basic understanding of kinetic theory might suggest, the molecules in liquid hydrogen are still very widely spaced apart. This can be seen from its density, which is only 71 kg.m⁻³. When bonded to the metal the molecules are actually closer together, though the material is obviously very much denser, at 5470 kg.m⁻³.

Once the reaction has completed, and all the hydrogen has been released, then the whole procedure can be repeated.

Usually several hundred charge/discharge cycles can be completed. However, rather like rechargeable batteries, these systems can be abused. For example, if the system is filled at high pressure, the charging reaction will proceed too fast, and the material will get too hot, and will be damaged. Also, the system is damaged by impurities in the hydrogen, so a high purity hydrogen should be used.

The raw hydride material, as listed in Table 7.10, cannot of course be used by itself. It has to be contained in a vessel. Although the hydrogen is not stored at pressure, the container must be able to withstand a reasonably high pressure, as it is likely to be filled from a high pressure supply, and allowance must be made for human error. For example, the unit shown in Figure 7.17 will be fully charged at a pressure of 3 bar, but the container can withstand 30 bar. The container will also need valves and connectors. Even taking all these into account impressive practical devices can be built. In Table 7.14 below gives details of the very small 20 SL holder for applications such as portable electronics equipment, manufactured by GfE Metalle und Materialien GMBH of Germany, and shown in Figure 7.17. The volumetric measure, mass of hydrogen per litre, is nearly as good as for LH_2, and the gravimetric measure is not a great deal worse than for compressed gas, and very much the same as for a small compressed cylinder.

Table 7.14 Details of a very small metal hydride hydrogen container suitable for portable electronics equipment

Mass of empty container	0.26 kg
Mass of hydrogen stored	0.0017 kg
Storage efficiency (% mass H_2)	0.65 %
Specific energy	0.26 kWh.kg^{-1}
Volume of tank (approx.)	0.06 L
Mass of H_2 per litre	0.028 kg.L^{-1}

Figure 7.17   Small metal hydride hydrogen store for fuel cells used with portable electronics equipment.

One of the main advantages of this method is its safety. The hydrogen is not stored at a significant pressure, and so cannot rapidly and dangerously discharge. Indeed, if the valve is damaged, or there is a leak in the system, the temperature of the container will fall, which will inhibit the release of the gas. The low pressure greatly simplifies the design of the fuel supply system. It thus has great promise for a very wide range of applications where small quantities of hydrogen are stored. It is also particularly suited to applications where weight is not a problem, but space is. A good example is fuel cell powered boats, where weight near the bottom of the boat is an advantage, and is often artificially added, but space is at a premium.

The disadvantages are particularly noticeable where larger quantities of hydrogen are to be stored, for example in vehicles. The specific energy is poor. Also, the problem of the heating during filling and cooling during release of hydrogen becomes more acute. Large systems have been tried for vehicles, and a typical refill time is about one hour for an approximately 5 kg tank. The other major disadvantage is that usually very high purity hydrogen must be used, otherwise the metals become contaminated, as they react irreversibly with the impurities.

7.7.6   Alkali metal hydrides

An alternative to the reversible metal hydrides are alkali metal hydrides which react with water to release hydrogen, and produce a metal hydroxide. Some of these are shown in Table 7.10. Bossel (1999) has described a system using calcium hydroxide, for which the reaction is:-

$$Ca\,H_2\;+\;2\,H_2O\;\rightarrow\;Ca\left(OH\right)_2\;+\;2\,H_2 \qquad\qquad [7.13]$$

It could be said that the hydrogen is being released from the water by the hydride.

Another method that is used commercially, under the trade name "Powerballs", is based on sodium hydride. These are supplied in the form of polyethylene coated spheres of about 3cm diameter. They are stored underwater, and cut in half when required to produce hydrogen. An integral unit holds the water, product sodium hydroxide, and a microprocessor controlled cutting mechanism that operates to ensure a continuous supply of hydrogen. In this case the reaction is:-

$$Na\,H\;+\;H_2O\;\rightarrow\;Na\,OH\;+\;H_2 \qquad\qquad [7.14]$$

This is a very simple way of producing hydrogen, and its energy density and specific energy can be as good or better than the other methods we have considered so far. Sodium is an abundant element, and so sodium hydride is not expensive. The main problems with these methods are:-

• The need to dispose of a corrosive and unpleasant mixture of hydroxide and water. In theory, this could be recycled to produce fresh hydride, but the logistics of this would be difficult.

- The fact that the hydroxide tends to attract and bind water molecules, which means that the volumes of water required tend to be considerably greater than equations 7.13 and 7.14 would imply.
- The energy required to manufacture and transport the hydride is greater than that released in the fuel cell.

A further point is that the method does not stand very good comparison with metal air batteries. If the user is prepared to use quantities of water, and is prepared to dispose of water/metal hydroxide mixtures, then systems such as the aluminium/air or magnesium/air battery are preferable. With a salt water electrolyte, an aluminium/air battery can operate at 0.8 volts at quite a high current density, producing three electrons for each aluminium atom. The electrode system is much cheaper and simpler than a fuel cell.

Nevertheless, the method compares quite well with the other systems in several respects. The figures in Table 7.15 below are calculated for a self-contained system capable of producing 1 kg of hydrogen, using the sodium hydride system. The equipment for containing the water and gas, and the cutting and control mechanism is assumed to weigh 5 kg. There is three times as much water as equation 7.14 would imply is needed.

Table 7.15 Figures for a self contained system producing 1 kg of hydrogen using water and sodium hydride

Mass of container and all materials	45 kg
Mass of hydrogen stored	1.0 kg
Storage efficiency (% mass H2)	2.2 %
Specific energy	$0.87 \ kWh.kg^{-1}$
Volume of tank (approx.)	50 L
Mass of H_2 per litre	$0.020 \ kg.L^{-1}$

The storage efficiency compares well with other systems. This method may well have some niche applications where the disposal of the hydroxide is not a problem, though these are liable to be limited.

7.7.7   Storage methods compared

Before comparing the different storage methods we must make a link back to the earlier sections of this chapter, and make a comparison with methanol. As was said in Section 7.5, this is comparatively easily reformed to hydrogen, and makes a very promising store of hydrogen, even for quite small systems.

Methanol can be reformed to hydrogen by steam reforming, according to the following reaction:-

$$CH_3OH + H_2O \rightarrow CO_2 + 3H_2$$

This would yield 0.188 kg of hydrogen for each kg of methanol. However, full utilisation is not possible – it never is with gas mixtures containing carbon dioxide. Also, some of the product hydrogen is needed to provide energy for the reforming reaction. If we assume that the hydrogen utilisation can be 75%, then we can obtain 0.14 kg of hydrogen for each kg of methanol. As we have seen, a particularly pertinent application of methanol derived hydrogen is the case of motor vehicles. We can speculate that a 40 litre tank of methanol might be used, with a reformer of about the same size and weight as the tank. Such a system should be possible in the reasonably near term, and would give the following figures:-

Table 7.16 Speculative data for a hydrogen source, storing 40 L of methanol

Mass of reformer and tank	31.6 kg
Mass of hydrogen stored[a]	4.4 kg
Storage efficiency (% mass H2)	13.9 %
Specific energy	5.5 kWh.kg^{-1}
Volume of tank (approx.)	0.08 m^3
Mass of H$_2$ per litre	0.055 kg.L^{-1}

a) Assuming 75% conversion of available H$_2$ to usable H$_2$

It can be seen that this compares very favourably with all the other methods, including cryogenic hydrogen. Furthermore, methanol is greatly safer, simpler and cheaper to transport.

Table 7.17 below shows the range of gravimetric and volumetric measures for the main candidate systems for which data is currently available. With the exception of methanol, which is included for comparison, this data refers to real systems that are available, and not to any material or method that may or may not be available in the future.

Table 7.17 Data for comparing methods of storing hydrogen fuel. The figures refer to complete working systems, not the materials on their own.

Method	Gravimetric Storage efficiency, % mass hydrogen	Volumetric Mass (in kg) of hydrogen per litre
Pressurised gas	0.7 – 3.0	0.015
Reversible metal hydride	0.65	0.028
Water + metal hydride	2.2	0.020
Cryogenic liquid	14.2	0.040
Reformed methanol	13.9	0.055

Obviously these figures cannot be used in isolation – they don't include cost for example. Safety aspects do not appear in this table either. Several of the systems are plainly not suitable for very small-scale applications. Another factor is that some of the systems are also a method of hydrogen transport and supply, whereas others are not. For example, the alkali metal hydride and water method can be used to supply small users with hydrogen, as the compounds are reasonably low cost, and can easily be transported. The reversible metal hydride system is only a store, and must be refilled from a *local* source of hydrogen.

We are concerned here with storing hydrogen for use in fuel cells. It is therefore worth comparing these systems with rechargeable batteries, since in many cases this will be the alternative. A good example is the small metal hydride container shown in Figure 7.17, which holds 0.0017 kg of hydrogen. Using the figures given in Appendix 2, Section A 2.4, we can see that for a cell operating at $V_c = 0.6$ volts, which corresponds to an efficiency of 40% (HHV), the hydride container will delivery $26.6 \times 0.6 \times 0.0017 = 0.027$ ŀWh, or 27 Wh, of electrical energy. This is approximately the same as the capacity of 6 size D nickel cadmium cells, each of which has roughly the same volume as the hydrogen store. The cost of the hydride store is about the same as 6 size D NiCads, but the recharge time is only a few minutes. So we have obtained approximately six times the energy density, and a much faster recharge time, with no increase in cost.

Of course, this comparison is not fair, since we have ignored the fuel cell! The size of this will vary with the power required. In some applications the fuel cell might be about the same size again as the hydride store, so we would have a system of about three times the energy density of NiCad batteries. The key factor here is the ratio:-

$$\frac{Maximum\ power}{Energy\ required}$$

The lower this figure, the more attractive the fuel cell option is. However, in some cases a hybrid system is appropriate, as discussed in Chapter 9, Section 9.5.

Fuel processing and hydrogen storage are vital aspects of fuel cell system design, but they are not the only sub-systems that need to be added to the fuel cell stack. In the next chapter we address the vitally important issues connected with moving the reactant gases through the system.

References

Baker, B.S. (1965) *"Hydrocarbon fuel cell technology."* Academic Press, New York and London.

Bossel U.G. (1999)"Portable fuel cell battery charger with integrated hydrogen generator", Proceedings of the European Fuel Cell Forum Portable Fuel Cells conference, Lucerne, pp 79 – 84

Bain A. & VanVorst WD.(1999) "The Hindenburg tragedy revisisted: the fatal flaw found", *International Journal of Hydrogen Engergy*, Vol.24, No.3, pp399-403

Brown D.J., Brown R.A., Cooke B.H., Haynes D.A., Taylor M.R., and Blyth Z., (1996) " A study assessing electricity generation integerating coal gasifiers and fuel cells," ETSU report F/02/00026/REP, AEA Technology, Harwell, UK.

Dicks A.L., (1996) "Hydrogen generation from natural gas for the fuel cell systems of tomorrow," *Journal of Power Sources* vol. 61 pp 113-124.

Edlund D.J., and Pledger, W.A., (1998) "Pure hydrogen production from a multi-fuel processor." Proceedings of the US Fuel Cell Seminar, Palm Springs California Nov 16-19.

Edwards N., Ellis S.R., Frost J.C., Golunski S.E., van Keulen A.N.J., Lindewald N.G., and Reinkingh J.G., (1998) "Onboard hydrogen generation for transport applications: The HotSpot™ methanol processor", *Journal of Power Sources*, 781, pp. 123-128.

Fischer L. (1999) "Hydrogen storage devices using metal hydrides", Proceedings of the European Fuel Cell Forum Portable Fuel Cells conference, Lucerne, pp171 – 180

Flot D.M., and Irvine J.T.S., (1999) "Doped Mg_2TiO_4 spinels as potential SOFC anode materials", Extended abstracts of the 12th International Conference on Solid State Ionics, June 6-12, Halkidiki, Greece, pp. 213-214.

Freni S., and Maggio G., (1997) *International Journal of Energy Research*, Vol 21. pp. 253-264.

*Hord J. (1978) "Is hydrogen a safe fuel?", *International Journal of Hydrogen Energy*, Vol.3, pp157-176

Kahrom H. (1999) "Clean hydrogen for portable fuel cells", Proceedings of the European Fuel Cell Forum Portable Fuel Cells conference, Lucerne, pp159 – 170

Kalhammer F.R., Prokopius P.R., Roan V and Voecks G.E., (1998) "Status and prospects of fuel cells as automobile engines," report prepared for the State of California Air Resources Board.

Kumar R., Ahmed S., and Krumpelt M., (1996) "The low temperature partial oxidation reforming of fuels for transportation fuel cell systems." Proceedings of the US Fuel Cell Seminar, Kissimee, Florida November 17-20.

Langnickel U., (1999) "Fuel cell using digester gas", *European Fuel Cell News* Vol.6 No. 2, p4

Ledjeff-Hey K., Kalk Th., Mahlendorf F., Niemzig O., and Roes J., (1999) " Hydrogen by cracking of propane." Proceedings of the International Conference on Portable Fuel Cells, Lucerne, 21-24 June (ISBN 3-905592-3-7) pp193-203.

Liebhafsky H.A., and Cairns E.J. (1968), *"Fuel Cells and Fuel Batteries."* John Wiley & Sons. Inc.

Okada O., Tabata T., Takami S., Hirao K., Masuda M., Futjita H., Iwasa N., and Okhama T., (1992) "Development of an advanced reforming system for fuel cells" Int. Gas Research Conf. , Orlando Florida, USA, 15-19 Nov.

*Pehr K. (1994) "Aspects of Safety and Acceptance of LH_2 tank systems in passenger cars", in *Hydrogen Energy Progress X* , pp 1399 –1404

Perry Murray E., Tsai T., and Barnett S.A. (1999) "A direct methane fuel cell with ceria-based anode", *Nature*, 400, pp 649-651.

*Reister D., & Strobl W. (1992) "Current development and outlook for the hydrogen fuelled car" in *Hydrogen Energy Progress IX*, pp 1202- 1215

Rostrup-Nielsen J.R. (1993) *Catalysis Today*, Vol. 18, pp305-324

Rostrup-Nielsen J.R. (1984) in J.R. Anderson and M. Boudart (eds.) "Catalytic Steam Reforming" *Catalytic Science and Engineering.* Vol.5, Springer Verlag

Smith J.G.(1996) in Williams K.R., (1996) *"An introduction to fuel cells"*, Elsevier Publishing Co. Chapter 6, pp.214-246.

Teagan W.P., Bentley J., and Barnett B., (1998) "Cost implications of fuel cells for transport applications: Fuel processing options," *Journal of Power Sources*,Vol 71, pp. 80-85.

Twigg M., (1989) *Catalyst Handbook*, 2nd edition Wolfe, London

Van Hook J.P., (1981) Catalysis Review. Sci-Eng. Vol 21 No. 1.

Warren D. et al. (1986) "Performance evaluation for the 40 kW fuel cell power plant utilizing a generic landfill gas feedstock", KTI corp report (Southern California Edison.

*Zieger J. (1994), "HYPASSE – Hydrogen powered automobiles using seasonal and weekly surplus of electricity", in *Hydrogen Energy Progress X*, pp1427 – 1437

*These papers are reprinted in Norbeck J.M. et. al. (1996) *Hydrogen Fuel for Surface Transportation*, pub. Society of Automotive Engineers.

8

Compressors, Turbines, Ejectors, Fans, Blowers and Pumps

8.1   Introduction

Air has to be moved around fuel cell systems for cooling, and to provide oxygen to the cathode. Fuel gas often has to be pumped around the anode side of the fuel cell too. To do this pumps, fans, compressors and blowers have to be used. In addition, the energy of exhaust gases from a fuel cell can sometimes be harnessed using a turbine, making use of what would otherwise go to waste. The technology for such equipment is very mature, having already been developed for other applications. The fuel cell system designer will be choosing devices principally designed for other markets and products. The needs of fuel cells vary very widely, as does their size and application, and so we need to look at a wide range of "gas moving devices". These will have been primarily designed for a wide field of application, from turbochargers in diesel engines to aerators for fish tanks!

It is hoped that the reader will find much useful information in this chapter. It has been put here after the descriptions of and discussions about the main types of fuel cell, as the content is relevant to all of them. The topics dealt with are:-

- The very important, and quite complex, subject of **compressors,** their different types and performance. This is covered in Sections 8.2 to 8.7, and includes the derivation of two very useful equations. The first gives the temperature rise of a gas as it is compressed (equation 8.4). The second gives the power needed to run an air compressor, equation 8.7.
- **Turbines,** which can sometimes be used to harness the energy of exhaust gases, are briefly covered in Sections 8.8 and 8.9.
- **Ejectors,** a very simple type of pump, that can often be used to circulate hydrogen gas if it comes from a high pressure store, or for recycling anode gases, are discussed in Section 8.10
- **Fans and blowers**, as used for cooling and for cathode gas supply in small fuel cells, are covered in Section 8.11.
- **Membrane or diaphragm pumps**, used to pump reactant air and hydrogen through small (200W) to medium (3kW) PEM fuel cells, are discussed briefly in Section 8.12.

8.2   Compressors – Types Used

The types of compressors used in fuel cell systems are the same as those used in other engines, especially diesel engines. The four main types are illustrated below in Figure 8.1.

The simplest to understand is the Roots compressor shown in Figure 8.1(a). It is easy to envisage how this will pump and compress any gas when the two elements rotate, pushing the gas through. The Roots compressor is quite cheap to produce, and it works over a wide range of flowrates. However, it only works at reasonable efficiencies when producing a small pressure difference. Until recently it had been regarded as rather the lower quality and cheaper type of compressor. However, of late the Eaton Corporation in the USA has revived the Roots compressor. Their device has the refinements that the rotors have been twisted by $60°$ to form a partial helix, and the counter-rotating rotors have three lobes, instead of the two of Figure 8.1(a). These improvements, while increasing the cost, also greatly reduce the pressure variations and hence the noise, and also increase the efficiency, though they are still only suitable for a small pressure change, boosting by a factor of up to 1.8. They are used as superchargers for petrol engines, and are fitted as standard on certain types of Ford cars (among others), as well as having a strong after-sales accessory market.

It is not so easy to see how the Lysholm or screw-compressor of Figure 8.1(b) works. There are two screws, as can be seen in the photograph of the model of Figure 8.2(b), which counter rotate, driving the gas up the region between the two screws, and compressing it at the same time. It can be thought of as a refinement of the "Archimedes Screw" that has been used for pumping water since ancient times. There are two variations of the screw compressor. In the first, an external motor drives only one rotor, and the second rotor is turned by the first. For this to work the rotors are in contact, and so must be lubricated with oil. This type of compressor is widely used to provide compressed air for pneumatic tools and other industrial application needs. The small amounts of oil that are inevitably carried out with the air do not matter. In the second type the two rotors are connected by a synchronising gear – a separate pair of cogs provides the driving link from one rotor to the other. The counter rotating screws do not come into contact, though of course for a good efficiency they will run very close to each other. This type of compressor gives an oil free output, and is the sort we need in a fuel cell system, though it also has other applications, for example as a compressor for refrigeration systems.

Among the advantages of the Lysholm compressor are that it can be designed to provide a wide range of compression ratios – the exit pressure can be up to eight times the input pressure. (This is done by changing the length and pitch of the screws.) Another advantage is that they work at a good efficiency over a wide range of flow rates. However, they are expensive to manufacture - the rotors are clearly precision items. The axles that they turn on are subject to both lateral and axial loading, which makes the bearings more complex and expensive, and the synchronising gears add further complication and expense.

The most common type of compressor is the centrifugal or radial type illustrated in Figure 8.1(c). The gas is drawn in at the centre and flung out at high speed to the outer volute. Here the kinetic energy is "converted" to a pressure increase. The radial compressor is used on the vast majority of engine turbocharging systems. The precise shape of the rotating vane has been developed after much research and experience. An example rotor is shown in Figure 8.2. Although the shape may be complex, it can usually be cast as one piece. This type of compressor is thus low cost, well developed, and available to suit a wide range of flowrates. The efficiency of the centrifugal compressor is as good as any other type, but it must be operated within quite well defined flowrates and pressure changes to obtain these efficiencies. Indeed, it cannot operate at all at low flowrates, as is explained in Section 8.6 below. A further problem is that the rotor must rotate at very high speed - 80,000 rpm being typical. (For comparison, screw and Roots compressors rotate at up to about 15,000 rpm.) This means that care must be taken with the lubrication of the bearings.

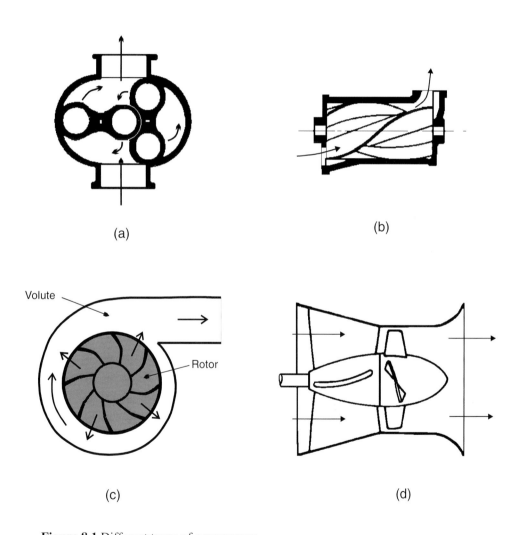

Figure 8.1 Different types of compressor

Figure 8.2 (a) Typical centrifugal air compressor rotor   **(b)** Model of twin screw Lysholm compressor

The axial flow compressor, as in Figure 8.1(d), works by driving the gas using a large number of blades rotating at high speed. It is, in essence, the inverse of the turbine commonly used in gas and steam thermal power systems. As with these steam or gas turbines, there must be as small as possible a gap between the ends of the blades and the housing. This means they are expensive to manufacture. The efficiency is good, but only over a fairly narrow range of flowrates. Experience with diesel systems suggests that the axial flow compressor will only be considered for systems above a few MW, operating at between half and full power at all times. (Watson, 1982)

A type of compressor that is used in some industrial air compressors is the rotating vane compressor. This claims advantages in terms of cost over the screw compressor. However, they cannot be used for fuel cells, since the tips of the rotating vanes *must* run over a film of oil, and so, even after filtering, there will always be some oil in the output gas, which is not generally acceptable for fuel cells.

Whatever type of compressor is being used, the symbol used in process flow diagrams is as shown below in Figure 8.3.

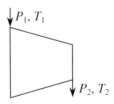

P_1, T_1

P_2, T_2

Figure 8.3 Symbol used for compressors

8.3   Compressor Efficiency

As with fuel cells, compressor efficiency is not easy to define. Whenever a gas is compressed, work is done on the gas, and so its temperature will rise, unless the compression is done very slowly or there is a lot of cooling. In a reversible process, which is also adiabatic (no heat loss) then it can readily be shown (using advanced high school or early university physics) that if the pressure changes from P_1 to P_2, then the temperature will change from T_1 to T_2', where:

$$\frac{T_2'}{T_1} = \left(\frac{P_2}{P_1}\right)^{\frac{\gamma - 1}{\gamma}}$$

[8.1]

This formula gives the constant entropy, or isentropic temperature change, as is indicated by the ' on T_2'. γ is the ratio of the specific heat capacities of the gas, C_p/C_v..

In practice the temperature change will be higher that this. Some of the motion of the moving blades and vanes serves only to raise the temperature of the gas. Also, some of the gas might "churn" around the compressor, doing little but getting hotter. We call the actual new temperature T_2.

To derive the efficiency, we use the ratio between the following two quantities:

- The actual work done to raise the pressure from P_1 to P_2
- The work that would have been done if the process had been reversible or isentropic – the isentropic work.

To find these two figures we make some assumptions that are generally valid:

- The heat flow from the compressor is negligible
- The kinetic energy of the gas as it flows into and out of the compressor is negligible, or at least the change is negligible
- The gas is a "perfect gas", and so the specific heat at constant pressure, C_P, is constant.

With these assumptions, the work done is simply the change in enthalpy of the gas:

$$W = c_p \ (T_2 - T_1)m$$

where m is the mass of gas compressed. The isentropic work done is:

$$W' = c_p \ (T_2' - T_1) \ m$$

The isentropic efficiency is the ratio of these two quantities. Note that the name "isentropic" does not mean we are saying the process *is* isentropic, rather we are *comparing it with* an isentropic process.

$$\eta = \frac{isentropic \ work}{real \ work} = \frac{c_p \ (T_2' - T_1)m}{c_p \ (T_2 - T_1)m}$$

and so $\qquad\qquad \eta_c = \frac{T_2' - T_1}{T_2 - T_1}$ [8.2]

If we substitute equation 8.1 for the isentropic second temperature we get:

$$\eta_c = \frac{T_1}{(T_2 - T_1)} \left(\left(\frac{P_2}{P_1} \right)^{\frac{\gamma - 1}{\gamma}} - 1 \right) \qquad [8.3]$$

In this derivation we have ignored the difference between the static and the stagnation temperatures. This is important when the gas velocity is high. However, in the fuel cell itself the kinetic energy of the gas is not important. Even in diesel engines, whose flowrates after the compressor are very unsteady, and often rapid, the equations given here are usually used.

It is useful to rearrange the equation to give the change in temperature:

$$\Delta T = T_2 - T_1 = \frac{T_1}{\eta_c} \left(\left(\frac{P_2}{P_1} \right)^{\frac{\gamma - 1}{\gamma}} - 1 \right) \qquad [8.4]$$

This definition of efficiency does not consider the work done on the shaft driving the compressor. To bring this in we should also consider the mechanical efficiency η_m, which takes into account the friction in the bearings, or between the rotors and the outer casing (if any). In the case of centrifugal and axial compressors this is very high, almost certainly over 98%. We can say then that:

$$\eta_T = \eta_m \cdot \eta_c \qquad [8.5]$$

However, it is the isentropic efficiency η_c that is the most useful, because it tells us how much the temperature rises.

The rise in temperature can be quite high. For example, using air at 20° C (293 K) for which $\gamma = 1.4$, a doubling of the pressure, and using a typical value for η_c of 0.6, equation 8.4 becomes:

$$\Delta T = \frac{293}{0.6} \left(2^{0.286} - 1 \right)$$

$$= 170 \text{ K}$$

For some fuel cells this rise in temperature is useful as it pre-heats the reactants. On the other hand, for low temperature fuel cells it means the compressed gas needs cooling. Such coolers between a compressor and the user of the gas are called "intercoolers". We need to know the temperature rise in any case, but equation 8.4 also allows us to find the power.

8.4   Compressor Power

The power needed to drive a compressor can be readily found from the change in temperature. If we take unit time, then clearly:

$$Power = \dot{W} = c_p \, . \, \Delta T \, . \, \dot{m}$$

where $\dot{m}$ is the rate of flow of gas, in kg.s^{-1}. ΔT is given by equation 8.4, and so we have:

$$Power = c_p \frac{T_1}{\eta_c} \left[\left(\frac{P_2}{P_1} \right)^{\frac{\gamma-1}{\gamma}} - 1 \right] . \dot{m} \qquad [8.6]$$

This is obviously a very useful general equation. In the case of an air compressor, which is a feature of many fuel cell systems, we use the values for air of:-

$$c_p = 1004 \ J.kg^{-1} \ K^{-1}$$

and

$$\gamma = 1.4$$

$$\text{So power} = \ 1004 \frac{T_1}{\eta_c} \left[\left(\frac{P_2}{P_1} \right)^{0.286} - 1 \right] \dot{m} \ \ Watts \qquad [8.7]$$

The isentropic efficiency η_c can be found, or at least estimated, from charts like those described in the section following. To find the power needed from the motor or turbine driving the compressor the figure found from equation 8.7 should be divided by the mechanical efficiency, η_m. For a good estimate, 0.9 can be used for this. However, when specifying a compressor from a supplier, the figure from equation 8.7 is best used unaltered.

8.5   Compressor Performance Charts

The efficiency and performance of a compressor will depend on many factors, including:

- Inlet pressure, P_1
- Outlet pressure, P_2
- Gas flowrate, $\dot{m}$
- Inlet temperature, T_1
- Compressor rotor speed, N
- Gas density, ρ

- Gas viscosity, μ

To try and tabulate or draw on some kind of map the compressor performance with all these variables would clearly be hopeless. It is necessary to eliminate or group together these variables. This is often done in the following way:

- The inlet and outlet pressure are combined into one variable, the pressure ratio P_2/P_1

- Since, for any gas, density $\rho = P/RT$, and P and T are being considered, the density can be ignored. It is considered via the inclusion of pressure and temperature.

- It turns out that the viscosity of the gas, bearing in mind the limited range of gases normally used, can be ignored.

Further simplification is done by a process of dimensional analysis, which can be found in texts on turbines and turbochargers, (e.g. Watson and Janota, 1982). The result is to group together variables in "non-dimensional" groups. They are not really non-dimensional, but that is because various constants, with dimensions, have been eliminated from the analysis.

$$\text{The two groups are} \quad \frac{\dot{m}\sqrt{T_1}}{P_1} \quad \text{and} \quad \frac{N}{\sqrt{T_1}}.$$

These are called the "mass flow factor" and the "rotational speed factor" respectively. Other texts call them the "non-dimensional" mass flow and rotational speed. Charts are then plotted of the efficiency, for different pressure ratios and mass flow factors. Lines are plotted on these charts of constant rotational speed factor. The chart for a typical screw compressor (Lysholm) is shown in Figure 8.4. The lines of constant efficiency are like the contours of a map. Instead of indicating hills, they indicate areas of higher operating efficiency.

The units generally used for P_1 is the bar, and for temperature we use Kelvin. We can relate the mass flow factor to the power of a fuel cell reasonably simply. If we assume typical fuel cell operating conditions, (i.e. the air stoichiometry $= 2$, and the average fuel cell voltage $= 0.6$ volts), then, using equation A2.4, the flow rate of the air for a 250 kW fuel cell is:-

$$\frac{3.57\times10^{-7}\times2\times250,000}{0.6} = 0.3 \quad \text{kg.s}^{-1}$$

If we then assume standard conditions for the air (i.e. $P_1 = 1$ bar, $T = 298$ K then the mass flow factor is:-

$$\frac{0.3\times\sqrt{298}}{1.0} = 5.18 \approx 5 \quad \text{kg.s}^{-1}\text{K}^{1/2}\text{bar}^{-1}$$

So, the horizontal x axis of Figure 8.4 corresponds to the air flow needs of fuel cells of power approximately 0 to 250 kW. Similarly, if the rotor speed factor is 1000, this will correspond to a speed of about 17,000 rpm. (Note, these units have to run fast, the centrifugal compressors considered in the next section have to turn even faster!)

The use of these "non-dimensional" quantities is standard practice in text-books on compressors and turbines. However, it has to be said that in many manufacturers' data sheets they are not used. Standard condition figures are applied ($P = 1.0$ bar, $T = 298$ K), and the mass flow factor is replaced with mass flow rate, or even volume flow rate, and the rotational speed factor is replaced with speed in rev.min^{-1}. Generally speaking such charts will give satisfactory results - except in the case of multistage compressors. When gas has been compressed through the first stage its temperature and pressure will obviously have changed markedly, and so the mass flow factor will be quite different, even though the actual mass flow rate is unchanged.

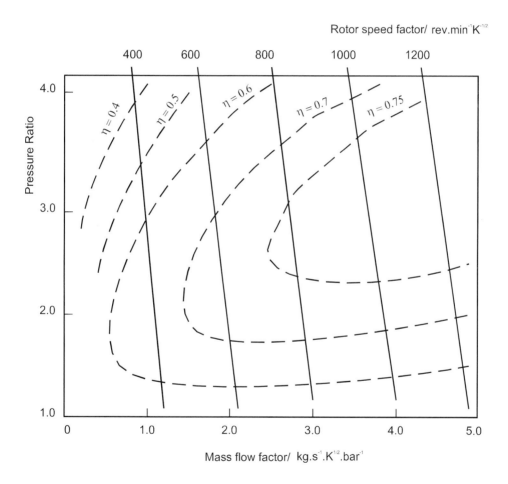

Figure 8.4 Performance chart for a typical Lysholm or Screw compressor. A mass flow factor of 5 corresponds to the air needs of a fuel cell of power about 250 kW.

Worked example.

A fuel cell stack with an output power of 100 kW operated at a pressure of 3 bar. Air is fed to the stack using the Lysholm compressor whose chart is shown in Figure 8.4. The inlet air is at 1.0 bar and 20 °C. The fuel cell is operated at an air stoichiometry of 2, and the average cell voltage in the stack is 0.65 volts, corresponding to an efficiency of 52% (ref. LHV). Use the chart and other equations to find:-

- The required rotational speed of the air compressor
- The efficiency of the compressor
- The temperature of the air as it leaves the compressor
- The power of the electric motor needed to drive the compressor.

Solution. First we find the mass flow rate of air using equation A2.4

$$\dot{m} = \frac{3.57 \times 10^{-7} \times 2 \times 100,000}{0.65} = 0.11 \quad \text{kg.s}^{-1}$$

This is then converted to the mass flow factor.

$$\text{Mass flow factor} = \frac{0.11 \times \sqrt{293}}{1.0} = 1.9 \quad \text{kg.s}^{-1}\text{K}^{1/2}\text{bar}^{-1}$$

We now find the speed and the efficiency of the compressor using the chart. Find the intercept of a horizontal line drawn from pressure ratio = 3, and a vertical line starting from the x axis at mass flow factor = 1.9. This will be very close to the 600 rotor speed factor line, and also the $\eta = 0.7$ "efficiency contour". So we can say that the rotor speed is:-

$$600 \times \sqrt{293} = 10300 \quad \text{rpm.}$$

The efficiency of the compressor is 0.7 or 70%. We can use this, and the mass flow rate, to find the temperature rise and the compressor power. The temperature rise is found using equation 8.4.

$$\Delta T = \frac{293}{0.7} \left(3^{0.286} - 1 \right) = 155 \quad \text{K}$$

Since the entry temperature is 20 °C, this means that the exit temperature is 175 °C. (So we note that if this is a PEM fuel cell it will need cooling. On the other hand, if it is a phosphoric acid fuel cell the preheating will already have been done.)

To find the compressor power we use equation 8.7

$$Power, \ \dot{W} = 1004 \times \frac{293}{0.7} \left(3^{0.286} - 1 \right) \times 0.11 = 17.1 \quad \text{kW}$$

This is the power of the compressor ignoring any mechanical losses in the bearings and driveshafts. The electric motor will not be 100% efficient either, so we could estimate that the power of the motor should be about 20 kW.

- Note that this 20kW of electrical power will have to come from the 100 kW fuel cell, consuming 20% of its output. This parasitic load is a major problem when running systems at pressure, and its importance for PEM fuel cells in discussed in Chapter 4.
- Note also that in this worked example we have assumed that we are working with air of fairly standard (i.e. low) water content. It was pointed out in Chapter 4 that the inlet air to PEM fuel cells is sometimes humidified. This has the effect of altering both the specific heat capacity, and γ, the ratio of the specific heat capacities, and this can alter the performance of the compressor. However, if the inlet air needs to be humidified, this is usually done after compression, when the air is hotter.

8.6 Performance Charts for Centrifugal Compressors

Centrifugal compressors are very common, being so inherently simple and reliable. However, they do have special problems, which merit careful consideration. These special features can be elucidated from their performance charts, an example being shown below in Figure 8.5. All centrifugal compressors have performance charts that are similar in form. The two main points to note are:

- There are regions of high efficiency, but they are very narrow. The constant efficiency "contours" are very close together when moving across the chart at a constant pressure ratio.
- There is a distinct "surge line". To the left of this line the compressor is unstable and should not be used. The centrifugal compressor works by accelerating gas out from the centre. If it does not have any gas "to work on", then it does not work, and the pressurised gas upstream of the pump will flow back, only to be pumped through again. The pressure thus becomes unstable and the gas gets hotter. The centrifugal pump must not be operated in this "low flowrate" region. If the flowrate has to be low, the pressure ratio must also be reduced. (This is a very simple explanation of 'surge' in centrifugal compressors; for a fuller explanation a book on turbochargers or compressors is recommended, e.g. Watson and Janota, 1982.)

A consequence of these two points, and the general shape of the performance chart, is that if a system using centrifugal compressors is to be operated at optimum efficiency, the pressure should *not* be constant. As the gas flowrate increases and decreases with varying power the pressure should also rise and fall, following the maximum efficiency regions of Figure 8.5. In most applications of centrifugal compressors this is not a problem. For example, when turbocharging an internal combustion engine, at lower engine speeds less pressure ratio, and hence supercharging, is expected. Similarly, with fuel cells less pressure boost is needed at lower powers. However, both the reactant air and fuel must be at a similar pressure, otherwise the stack may be damaged by pressure differentials, and the control needed to keep both pressures changing, yet the same as each other, can be difficult. Also, if a fuel processing system is involved, this can further complicate matters.

Note that a rotor speed factor of 4500, the value in the optimum operating region, corresponds to $4500 \times \sqrt{298} \approx 78{,}000$ rpm. These are very high speed devices.

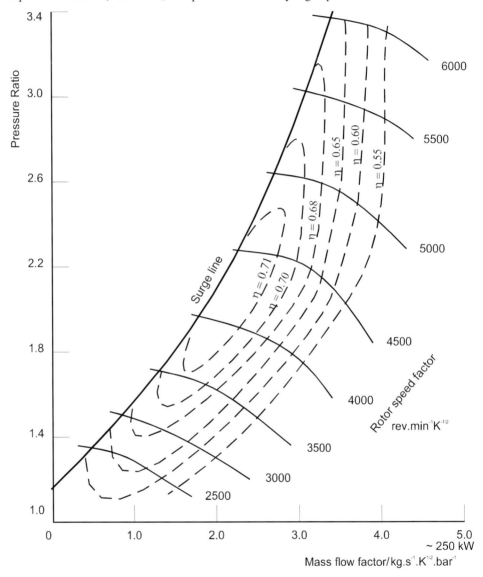

Figure 8.5 Performance chart for a typical centrifugal compressor

8.7   Compressor Selection – Practical Issues

It would be tempting to suppose the fuel cell system designer could select a compressor simply using the mass flow rate, calculated from the cell power using equation A2.4 in Appendix 2. Unfortunately this is unlikely to be the case, unless he is working on an unlimited budget! A system designer will normally have to design around what is actually available.

For pressure ratios in the range 1.4 to around 3, most of the compressors available are designed principally for use with internal combustion engines. A good example is the Eaton Supercharger, as mentioned in Section 8.2. This would be a good choice for a pressure ratio around 1.35 to 1.7 (or an outlet pressure of 0.35 to 0.7 bar gauge). However, such units are produced only for comparatively large petrol engines. The smallest of the range from this particular company is shown in Figure 8.6. It is designed for use in the flow range 50 – 100 litres.s^{-1}, corresponding to a fuel cell power range of about 50kW to 150 kW.[1] In terms of fuel cells, this is quite high power, and remember that this is for the *smallest* available. We can see then that obtaining such a compressor for fuel cells of power less than 50 kW is going to be difficult.

Figure 8.6 Eaton Supercharger. This unit is about 25cm long. It will boost pressure by between 35 and 70 kPa. (By kind permission, Eaton Corporation)

Figure 8.7 A Lysholm or screw compressor. The air enters a hole, not visible, on the left. It exits via the holes on the visible face. A pulley on the right drives the screws.

The same problem applies to the higher pressure ratios, of above 1.6 up to 3, which corresponds to 0.6 to 2 bar gauge, or about 9 to 30 psig. In this pressure range the screw compressor or Lysholm would be the first choice for efficiency and flexibility, as is shown by its chart in Figure 8.4. Such screw compressors are known commercially in the motor trade as "WhippleChargers", and the majority are manufactured by a company called Autorotor in Sweden. A typical unit is shown in Figure 8.7. The smallest unit in this range weighs just 5 kg, and its dimensions are 18 × 9 × 13 cm, so it is quite a small unit. However, it is designed for flow rates up to about 0.12 kg.s^{-1}, which corresponds to about 100kW for typical fuel cell operating conditions ($\lambda = 2$, cell voltage = 0.6). It can be seen from Figure 8.4 that when such units are operated at low flow rates, their efficiency falls off. So we can see, again, that we have a problem if we are looking for a compressor for small fuel cells. However, for fuel cells of 50 kW or more, and especially above 100 kW, there is a wide range of available products, where the designer can choose from well tried

---

[1] This was calculated using the equation A2.4 derived in Appendix 2 with typical values of stoichiometry ($\lambda = 2$) and average cell voltage ($V_c = 0.6$).

and tested units that are produced in quantity and so are readily available at reasonable cost.

The units shown in Figures 8.6 and 8.7 are designed for use on internal combustion engines. They both have pulleys for connection to the crankshaft of the engine. In fuel cell systems the compressor will have to be driven by an electric motor. Typically this will be as large or larger than the compressor, but motors are covered in more detail in the next chapter.

8.8   Turbines

Fuel cells usually output warm or hot gases. Fuel reformers do likewise. This energy can be harnessed with a turbine. The turbine will frequently be used to turn a compressor, to compress the incoming air or fuel gas. In a few cases there may be excess power, and in such cases a generator can be fitted to the same shaft.

There are only two types of turbine worth considering. The first is the centripetal or radial, which is essentially the inverse of the centrifugal compressor considered above. This is the preferred choice unless the powers concerned are over about 500 kW, when the axial turbine used as standard in gas and steam turbine power generation sets may be considered. Whatever type of turbine is used, the symbol is as shown below in Figure 8.8:

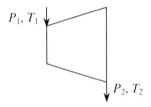

Figure 8.8 Symbol for a turbine.

The efficiency of the turbine is treated in a similar way to that for compressors, with the same assumptions. If the turbine worked isentropically then the outlet temperature would fall to T_2' where, as with the compressor:

$$\frac{T_2'}{T_1} = \left(\frac{P_2}{P_1} \right)^{\frac{\gamma-1}{\gamma}}$$ [8.1] (repeated)

However, because some of the energy will not be output to the turbine shaft, but will stay with the gas, the actual output temperature will be higher that this. The actual work done will therefore be *less* than the isentropic work. (For the compressor it was *more*). We thus define the isentropic efficiency as:

$$\eta_c = \frac{actual\ work\ done}{isentropic\ work}$$

Making the same assumptions about ideal gasses and so on that we did for compressors, this becomes:

$$\eta_c = \frac{T_1 - T_2}{T_1 - T_2'}$$

$$= \frac{T_1 - T_2}{T_1 \left(1 - \left(\dfrac{P_2}{P_1} \right)^{\frac{\gamma - 1}{\gamma}} \right)} \tag{8.8}$$

This is usually used to find the temperature change. So we can arrange it to:

$$\Delta T = T_2 - T_1 = \eta_c \cdot T_1 \left(\left(\frac{P_2}{P_1} \right)^{\frac{\gamma - 1}{\gamma}} - 1 \right) \tag{8.9}$$

Note that because $P_2 < P_1$ this will always be negative. It is useful to use this equation to derive a formula for the power available from the turbine. Applying the same reasoning and simplifying assumptions that we did in Section 8.4, we have:

$$Power = \dot{W} = c_p \cdot \Delta T \cdot \dot{m}$$

So

$$Power, \ \dot{W} = c_p \, \eta_c \, T_1 \left(\left(\frac{P_2}{P_1} \right)^{\frac{\gamma - 1}{\gamma}} - 1 \right) \dot{m} \tag{8.10}$$

To get the power available to drive an external load we should multiply this by η_m, the mechanical efficiency. As with the compressor, this should be about 0.98 or better.

The representation of turbine performance using charts is done along the same lines as for compressors. The performance map is drawn in a similar way, except that the vertical axis is now P_2/P_1, instead of P_1/P_2. The form of the map has similarities, though the shape and direction of the rotational speed factor lines are completely different. An example is shown in Figure 8.9 overleaf. For any given speed the mass flowrate rises as the pressure drop increases, as we would expect, but tends towards a maximum value. This maximum value is called the "choking limit". Naturally enough, this depends largely on the diameter of the turbine housing.

Notice that if you pick on a constant mass flow (e.g. 1.5), and move up through the map, the rotor speed factor increases as the pressure ratio increases. This again is what would be expected.

Worked example. What power would be available from the exit gases of the 100 kW fuel cell system given in Section 8.5?

Solution. The mass of the cathode exit gas will have been increased by the presence of water produced in the cells, but since this is the result of replacing O_2 with $2H_2O$, the mass change will be very small - hydrogen being so light[2]. So we will approximate $\dot{m}$, the mass flowrate as still 0.11 kg.s^{-1}. We can only estimate the exit temperature as, say 90°C (363 K), a typical operating temperature for this type of fuel cell. The entry pressure is 3.0 bar. The exit pressure must be a little less than this, say 2.8 bar. So can calculate the mass flow factor as:-

$$\text{Mass flow factor } \frac{0.11 \times \sqrt{363}}{2.8} = 0.75 \text{ kg.s}^{-1}.\text{K}^{1/2}.\text{bar}^{-1}$$

Using the turbine performance chart in Figure 8.9 we can find the efficiency and speed of the turbine. If we find the intercept up from 0.75 on the x axis and 2.8 on the Pressure ratio axis, we see that it is very close to the rotor speed factor = 5000 line, and in the $\eta = 0.70$ efficiency region. (More or less optimum operating conditions in fact!) So we can conclude that:-

$$\text{Rotor speed} = 5000 \times \sqrt{363} \approx 95,000 \text{ rpm}$$

Note that this is very fast! Such turbines are high speed devices. They could NOT be used to drive a screw compressor, though such speeds are suitable for driving centrifugal compressors, as is discussed in Section 8.9 below.

Since we know that the efficiency is about 0.7, we can use equation 8.10 to find the power. We are using air, for which c_p is 1004 J.kg^{-1}K^{-1}, and $\gamma = 1.4$, and T_1 is 363 K, so equation 8.10 becomes:-

$$\text{Power available} = 1004 \times 0.7 \times 363 \left(2.8^{0.286} - 1\right) \times 0.11 \approx 9.6 \text{ kW}.$$

This is a useful addition to the 100 kW electrical power output of the fuel cell, but note that it only provides about half of the power needed to drive the compressor, as calculated with the worked example of Section 8.5. Furthermore, this example is the best possible result, turbine efficiencies will usually be somewhat lower than the 0.7 we obtained here. As can be seen from the chart, much of the operating region is at greatly lower efficiencies.

8.9   Turbochargers

The turbine of the worked example in Section 8.8 might be used to drive a compressor. When this is done the turbine and compressor are mounted side by side on the same shaft. With the surrounding housing this makes a very compact and simple unit. They are usually called "turbochargers" because their main application is the supercharging of

---

[2] The addition of water to the air will raise the specific heat capacity c_p and lower γ, but the effect of this will not be great. We are only *estimating* the power.

engines using a turbine, driven by the exhaust gases. Turbocompressors might be a better name. The symbol for such a unit is shown below in Figure 8.10.

The technology of these units is very mature, and they are available in a range of sizes, but only for the larger fuel cells, 50 kW or more. They are used on the majority of modern diesel engines, and so are mass-produced. Compared to many other parts of a fuel cell system, they are readily available at low cost.

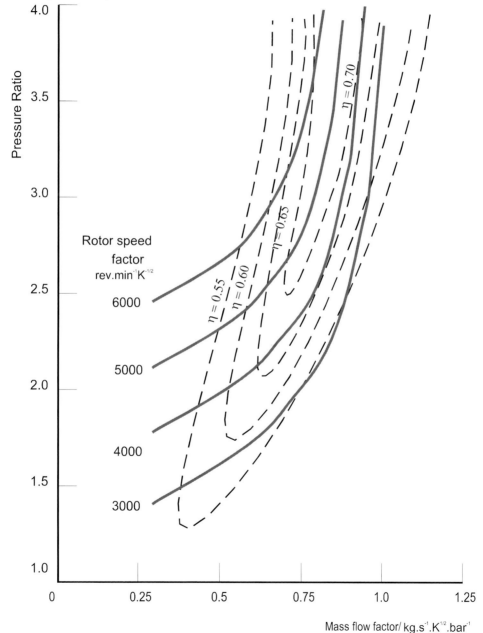

Figure 8.9 Performance chart for a typical small radial turbine

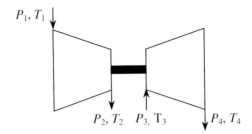

P_1, T_1

P_2, T_2 P_3, T_3 P_4, T_4

Figure 8.10 Representation of a linked turbine and compressor or "turbocharger"

8.10 Ejector Circulators

The ejector is the simplest of all types of pump. It has no moving parts. They are widely used in hydrogen fuel cells where the hydrogen is stored at pressure. In some Siemens-Westinghouse SOFCs they are used to recirculate the fuel gas. They harness the stored mechanical energy in the gas to circulate the fuel around the cell.

A simple ejector is shown in Figure 8.11. A gas or liquid passes through the narrow pipe A into the venturi B. It acquires a high velocity at B, and hence produces a suction in pipe C. The fluid coming through at A thus entrains with it the fluid from C and sends it out at D. Clearly, for this to work, the fluid from A must be at a higher pressure than that in C/B/D, otherwise it would not eject from pipe A at high velocity.

The fluid entering at A does not have to be the same as that in pipe C/B/D. The most common use of the ejector is in steam systems, with steam being the fluid passing through the narrow pipe and jet A. So, ejectors are used to pump air, for example to maintain vacuum in the condensers of steam turbines. They are also used to pump water into boilers, and since the steam mixes with the pumped water it also pre-heats it. The steam is also used to circulate lower pressure steam in steam heating systems. In this application it is closest to its normal use in fuel cell systems.

In fuel cell systems in which, for example, the hydrogen fuel is circulated by an ejector as in Figure 8.11, the gas is supplied at high pressure at A, and in the ejector the energy of the expanding gas draws the gas from the fuel cell at C, and sends it on through D back to the fuel cell. The pressure difference generated will be enough to drive the gas through the cell as well as through any humidifying equipment.

In terms of converting the mechanical energy stored in the compressed gas into kinetic energy of the circulating gas the ejector circulator is very inefficient. However, since the mechanical energy of the stored gas is not at all convenient to harness, it is essentially "free", and so long as the circulation is sufficient, the inefficiency does not matter. The internal diameter of pipe A, and the mixing region B, and C/D have to be chosen bearing in mind the pressure differences and the flowrates. Tables and graphs for these can be found in chemical engineering reference books such as Perry (1984).

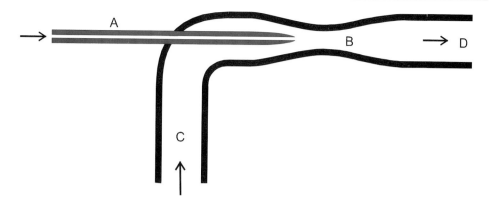

Figure 8.11 Diagram of a simple ejector circulation pump

8.11   Fans and Blowers

For straightforward cooling purposes fans and blowers are available in a huge range of sizes. The normal axial fan, such as we see cooling all types of electronic equipment, is an excellent device for moving air, but only against very small pressures. A small fan, such as is commonly used in electronic equipment, might move air at the rate of about 0.1 $kg.s^{-1}$, which is equivalent to 85 $L.s^{-1}$ at standard conditions. However, this flow rate will drop to zero if the back pressure even rises to 50 Pa. These pressures are so low, that they are often converted to cm of water and this helps us visualise how low they are - in this case just 0.5 cm of water! This is a typical maximum back pressure for such a fan. The result is they can only be used when the air movement is in a very open area, such as equipment cabinets, and a few very open designs of PEM fuel cell.

For a somewhat greater pressure the centrifugal fan, such as we often see being used to blow the air in heating units, would be used. These draw air in, and throw it out sideways. They are not totally different from the centrifugal compressors already considered, but turn at much slower speeds (by a factor of several hundred), have much longer blades, and a much more open construction. There are two main types, the 'backward curved' and

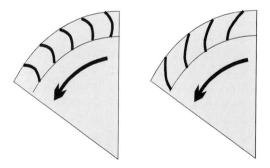

Figure 8.12   Forward curved and backward curved impeller
blades on a centrifugal blower.

'forward curved' centrifugal blowers. These names relate to the shape of the blades, as is shown in Figure 8.12 opposite. The forward curved blade gives a higher through put of air, but with a lower maximum back pressure - around 3 cm of water being a typical value. This type of blower is used to drive the cooling air through small to medium sized PEM fuel cells. The backward curved impeller is more suitable to cases where a higher pressure is needed, but we accept a lower flowrate. However, in this context "higher pressure" means a 3 to 10 cm of water - still virtually nothing! This type of blower is used for feeding air to furnaces, and might well be suitable for blowing the reactant air to small and medium sized PEM fuel cells. However, sometimes the pressure requirements might be greater, and a pump such as that described in the following section could be used.

With these fans and blowers the concept of efficiency is a difficult one. The input power is simply the electrical power used to run the motor. The purpose of the blower could be said to move the air, and so the output power is the rate of change of kinetic energy of the air. Using this measure fans and blowers usually have very low efficiencies which falls even lower as the back pressure against the air flow rises. However, this is not really a very helpful measure, since we do not actually want kinetic energy in the air. We usually want something else, for example the removal of heat. Let us take as an example a small 120 mm axial fan such as is often used to cool electronic equipment. Such a fan might move air at 0.084 kg.s^{-1}, and consume 15 watts of electric power. If the air it blows rises in temperature by just 10 °C, then the rate of removal of heat energy will be:-

Power $= C_p \times \Delta T \times \dot{m} = 1004 \times 10 \times 0.084 = 843$ Watts

That is 843 Watts of heat removed for just 15 Watts of electrical power. Is that efficient? It may be. But it is not if more of that heat could be removed by natural convection and radiation, and we are using a larger fan than necessary. The point is that the efficiency of the heat removal system depends more on the design of the air flow path, and how much temperature the air can pick up as it goes through, than on the performance of the motor. Rather than the efficiency of such a cooling system, it is more useful to define *effectiveness* as:-

$$\text{Cooling system effectiveness} = \frac{\text{rate of heat removal}}{\text{electrical power consumed}}$$

In this case, that would give a figure of 843/15 = 56. In the example system given in section 4.8, page 101, a centrifugal blower, probably forward curved, consumes 70 W of electrical power to dispose of 1.84 kW of heat. The effectiveness of this system is thus 26, somewhat lower than the previous case. That is because the rate of movement of air with a centrifugal blower is less than with an axial fan, and is the price that has to be paid for a more compact fuel cell, where it is more difficult to get the air through. On the other hand, the more compact systems should be able to achieve a greater temperature change in the air as it goes through. Generally, fuel cell system designers should obtain effectiveness figures of between 20 and 30 fairly readily. There is always a balance in cooling systems between flowrate of air and electrical power consumed. Higher flowrates improve heat transfer, but at the expense of fan power consumption.

8.12   Membrane/Diaphragm Pumps

Small to medium sized PEM fuel cells are liable to be an important market for fuel cells, in portable power systems. A problem with these cells is the pumps for circulating the reactant air, and the hydrogen fuel, if not dead-ended. In compact designs of, say, 200 W to 2 kW, the back pressure of the reactant air is liable to be around 10 kPa, equivalent to about 1m of water, or 0.1 bar. This is too great for the blowers and fans we have just been considering. On the other hand the flowrates are too low for the commercial blowers and compressors we considered in the earlier sections of this chapter. Another type of pump is needed. The main features of these pumps should be:-

- Low cost
- Silent
- Reliable when operated continuously for long periods
- Able to operate against pressures of about 10 kPa (1m water)
- Available in a suitable range of sizes, i.e. about 2.5×10^{-4} to 2.5×10^{-3} kg.s^{-1} (or 12 to 120 SLM)
- Efficient, low power consumption

The types of already existing pumps that satisfy a good deal of these requirements are made for applications such as gas sampling equipment, small scale chemical processing, and fish tank aerators. They are either vane or diaphragm pumps. It is the fish tank aerator pumps that "hit" most of the requirements above. These are mass produced, so low cost. They have to be silent, and are meant for continuous operation day after day for several years. The larger versions are also designed for operating against a metre or two of water, and so offer exactly the pressure range required. These pumps are diaphragm pumps, as in Figure 8.13 overleaf. There are many types of diaphragm (or membrane) pumps, since they have been used since ancient times. The diaphragm is moved up and down, shifting the air through the system by way of the two valves in a fairly obvious mechanical way. The use of the soft rubber makes for very quiet[3] and long-term reliable operation, though it does limit the operating pressure, but in this regard it is still suitable for our 10 – 20 kPa needs. Unfortunately it is not possible to use this type of pump directly "as sold", since the actuators they are usually supplied with are only suitable for AC operation, and not particularly efficient. However, the basic pump module, as in Figure 8.13, can be purchased very readily for a few dollars. The mass produced units will shift air at only about 5×10^{-4} kg.s^{-1}, but they can be readily operated in parallel. It is a fairly straightforward task to add a DC operated actuator or motor to the pump, or to several pumps. Furthermore, the flowrate can easily be controlled by modulating the force applied by the actuator.

An example of a published system design that uses this approach is for a 300 watt PEM fuel cell (Popelis et al., 1999). Using equation A2.4, and a stoichiometry of 2, and operation at about 0.6 volts per cell, a 300 Watt cell would require a reactant air flow rate

---

[3] In some systems the diaphragm pump is quite noisy. This is usually the result of imperfect match between the motor and the actuator, or noise in the mechanism that converts the circular motor action into linear motion for the pump.

of 3.57×10^{-4} kg.s^{-1}, which is equivalent to 18 SLM. The published details give a flow rate of 15 – 20 SLM.

This flow rate is delivered at a pressure of between 1.10 and 1.15 bar. The pump unit is manufactured by KNF Neuberger, and is driven by a 12 volt DC motor. According to the published data, this motor/pump combination consumes between 14 and 19 watts. The parasitic power loss is thus about 6%. This represents about as low a figure as can be expected for a small fuel cells system. For example, the 2 kW system presented in Section 4.9 used 200 watts of electrical power to provide the reactant air, a 10% parasitic loss.

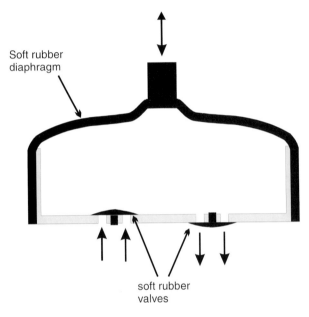

Soft rubber
diaphragm

soft rubber
valves

Figure 8.13   Diaphragm pump - low cost, quiet, and long-term reliable.

References

Daly B.B. (1979) *Woods Practical Guide to Fan Engineering*, Woods of Colchester Ltd.
Perry R.H. & Green D.W.,(1984) *Perry's Chemical Engineer's Handbook*, MrGraw-Hill, p.6-33.
Popelis I., Tsukada A, & Scherer G.,(1999) "12 Volt 300 Watt PEFC power pack", Proceedings of the European Fuel Cell Forum Portable Fuel Cells Conference, Lucerne, pp147-155
Watson N & Janota M.S., (1982) *Turbocharging the Internal Combustion Engine*, Macmillan

9

Delivering Fuel Cell Power

9.1   Introduction

We have covered the details of fuel cell performance, how the different types work, how they can be fuelled, and how the gases can be blown and compressed. In this final chapter we think about the basic purpose of a fuel cell - to generate electrical power. In this chapter we have brought together those areas of power electronics and electrical engineering that are of particular relevance to fuel cell systems, and hopefully explained them in a way that engineers from all specialities can understand. Fuel cells bring a few special problems, but by and large standard equipment and methods used in other electrical power systems can be employed. In other words, unlike many aspects of fuel cell systems, the electrical problems can generally be solved using more-or-less standard technologies. There are four technologies that we particularly need to address.

1. *Regulation.* The electrical output power of a fuel cell will often not be at a suitable voltage, and certainly that voltage will not be constant. We have seen in Chapter 3 that a graph of voltage against current for a fuel cell is by no means flat. Increasing the current causes the voltage to fall in all electrical power generators, but in fuel cells the fall is much greater. Voltage regulators, DC/DC converters, and chopper circuits are used to control and shift the fuel cell voltage to a fixed value, which can be higher or lower than the operating voltage of the fuel cell. These circuits are covered in Section 9.2.

2. *Inverters.* Fuel cells generate their electricity as direct current, DC. In many cases, especially for small systems, this is an advantage. However, in larger systems that connect to the mains grid, this DC must be converted to alternating current, AC. The inverters that do this are described and discussed in Section 9.3. Another important topic discussed in this section is that of the legislative framework that applies to generators connected to the mains.

3. *Electric Motors.* A major application area for fuel cells is transport, where the electrical power is delivered to an electric motor. In addition, motors are often needed to drive compressors and pumps within a fuel cell system. Most fuel cell systems of power greater than about one kilowatt will have at least three electric motors. The advent of cheaper power electronic devices and controllers has led to the development of new types of motor, with very high efficiencies. The competing claims of these different motor types are often hard to evaluate. In Section 9.4 we give an overview of these modern electric motors and their controllers, with a particular view to their application

within fuel cell systems. This includes an indication of the size and mass a motor of any given power is likely to have.

4. *Hybrid systems.* In terms of cost per watt of power fuel cells are expensive, and likely to remain so for some time to come. In many applications, for example transport and communications equipment, the average power needed is much lower than the peak power. In these cases lower cost systems can be designed using a battery (or capacitor) with a fuel cell in a hybrid system. Basically, the fuel cell runs at the average power, with the battery or capacitor providing the peak power. These hybrid systems are described in Section 9.5.

Finally, in Section 9.6, we look at an example of a complete fuel cell system in order to illustrate the issues discussed, both in this chapter, and Chapter 8.

9.2   DC Regulation and Voltage Conversion

9.2.1   Switching devices

The voltage from all sources of electrical power varies with time, temperature, and many other factors, especially current. However, fuel cells are particularly badly regulated. In Chapter 3, Figures 3.1 and 3.2, we see that the voltage from a cell falls rapidly with rising current density. Figure 9.1 summarises some data from a real 250 kW fuel cell used to drive a bus (Speigel et al., 1999). The voltage varies from about 400 to over 750 Volts, and we also see that the voltage can have different values at the same current. This is because, as well as current, the voltage also depends on temperature, air pressure, and on whether or not the compressor has got up to speed, among other factors.

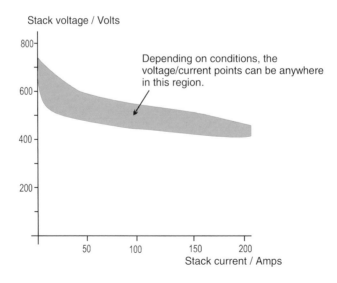

Figure 9.1 Graph summarising some data from a real 250 kW fuel cell used to power a bus. (Derived from data in Spiegel et al., 1999)

Most electronic and electrical equipment requires a fairly constant voltage. This can be achieved by dropping the voltage down to a fixed value below the operating range of the fuel cell, or boosting it up to a fixed value. This is done using "switching" or "chopping" circuits, which are described below. These circuits, as well as the inverters and motor controllers to be described in later sections use *electronic switches*.

As far as the user is concerned, the particular type of electronic switch used does not matter greatly, but we should briefly describe the main types used, so that the reader has some understanding of their advantages and disadvantages. Table 9.1 shows the main characteristics of the most commonly used types.

Table 9.1 Key data for the main types of electronic switch used in modern power electronic equipment.

Type	Thyristor	MOSFET	IGBT
symbol			
Max. voltage (V)	4500	1000	1700
Max. current (A)	4000	50	600
Switching time (µs)	10 - 25	0.3 – 0.5	1 – 4

The metal oxide semiconductor field effect transistor (MOSFET) is turned on by applying a voltage, usually between 5 and 10 volts, to the gate. When 'on' the resistance between the drain (d) and source (s) is very low. The power required to ensure a very low resistance is small, as the current into the gate is low. However, the gate does have a considerable capacitance, so special drive circuits are usually used. The current path behaves like a resistor, whose ON value is $R_{DS_{ON}}$. The value of $R_{DS_{ON}}$ for a MOSFET used in voltage regulation circuits can be as low as about 0.01 Ohms. However, such low values are only possible with devices that can switch low voltages, in the region of 50 Volts. Devices which can switch higher voltages have values of $R_{DS_{ON}}$ of about 0.1 Ohms, which causes higher losses. MOSFETs are widely used in low voltage systems of power less than about 1kW.

The insulated gate bipolar transistor, IGBT, is essentially an integrated circuit combining a conventional bipolar transistor and a MOSFET, and has the advantages of both. They require a fairly low voltage, with negligible current, at the gate to turn on. The main current flow is from the collector to the emitter, and this path has the characteristics of a p-n junction. This means that the voltage does not rise much above 0.6 volts at all current within the rating of the device. This makes it the preferred choice for systems where the current is greater than about 50 amps. They can also be made to withstand higher voltages. The longer switching times compared to the MOSFET, as given in Table 9.1, are a disadvantage in lower power systems. However, the IGBT is now almost universally the electronic switch of choice in systems from 1kW up to several hundred kW, with the 'upper' limit rising each year.

The thyristor has been the electronic switch most commonly used in power electronics. Unlike the MOSFET and IGBT the thyristor can only be used as an electronic switch – it has no other applications. The transition from the blocking to the conducting state is triggered by a pulse of current into the gate. The device then remains in the conducting state until the current flowing through it falls to zero. This feature makes them particularly useful in circuits for rectifying AC, where they are still widely used. However, various variants of the thyristor, particularly the gate-turn-off, or GTO thyristor, can be switched off, even while a current is flowing, by the application of a negative current pulse to the gate.

Despite the fact that the switching is achieved by just a pulse of current, the energy needed to effect the switching is much greater than for the MOSFET or the IGBT. Furthermore, the switching times are markedly longer. The only advantage of the thyristor (in its various forms) for DC switching is that higher currents and voltages can be switched. However, the maximum power of IGBTs is now so high, that this is very unlikely to be an issue in fuel cell systems, which are usually below 1MW in power.

Ultimately the component used for the electronic switch is not of great importance. As a result the circuit symbol used is often the "device independent" symbol shown in Figure 9.2. In use, it is essential that the switch moves as quickly as possible from the conducting to the blocking state, or vice-versa. No energy is dissipated in the switch while it is open circuit, and only very little when it is fully on; it is while the transition takes place that the product of voltage and current is non-zero, and that power is lost.

Figure 9.2 Circuit symbol for a voltage operated electronic switch of any type.

9.2.2   Switching regulators

The 'step-down' or 'buck' switching regulator (or chopper) is shown in Figure 9.3 opposite. The essential components are an electronic switch with an associated drive circuit, a diode, and an inductor. In Figure 9.3(a) the switch is on, and the current flows through the inductor and the load. The inductor produces a back EMF, making the current gradually rise. The switch is then turned off. The stored energy in the inductor keeps the current flowing through the load, using the diode, as in Figure 9.3(b). The different currents flowing during each part of this on-off cycle are shown in Figure 9.4. The voltage across the load can be further smoothed using capacitors if needed.

If V_1 is the supply voltage, and the 'on' and 'off' times for the electronic switch are t_{ON} and t_{OFF}, then it can be shown that the output voltage V_2 is given by:

$$V_2 = \frac{t_{ON}}{t_{ON} + t_{OFF}} V_1 \qquad\qquad [9.1]$$

It is also clear that the ripple depends on the frequency - higher frequency, less ripple. However, each turn-on and turn-off involves the loss of some energy, so the frequency should not be too high. A control circuit is needed to adjust t_{ON} to achieve the desired output voltage – such circuits are readily available from many manufacturers.

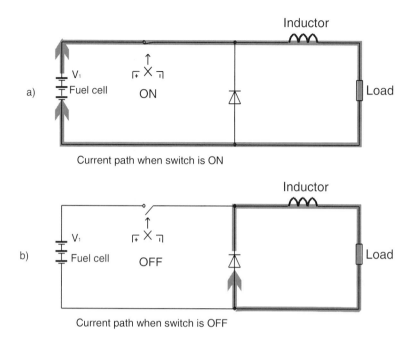

Figure 9.3 Circuit diagram showing the operation of a switch mode step down regulator.

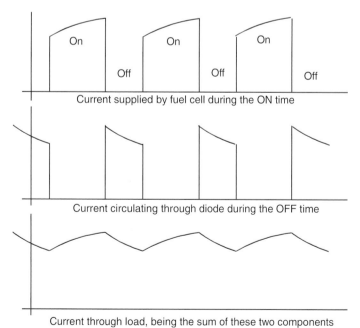

Figure 9.4 Currents in the step down switch mode regulator circuit

The main energy losses in the step-down chopper circuit are:

- Switching losses in the electronic switch
- Power lost in the switch while on, ($0.6 \times I$ for an IGBT or $R_{DS_{ON}} \times I^2$ for a MOSFET)
- Power lost due to the resistance of the inductor
- Losses in the diode, $0.6 \times I$

In practice all these can be made very low. The efficiency of such a step-down chopper circuit should be over 90%. In higher voltage systems, about 100 volts or more, efficiencies as high as 98% are possible.

We should at this point briefly mention the 'linear' regulator circuit. The principle is shown in Figure 9.5. A transistor is used again, but this time it is not switched fully on or fully off. Rather, the gate voltage is adjusted so that its resistance is at the correct value to drop the voltage to the desired value. This resistance will vary continuously, depending on the load current and the supply voltage. This type of circuit is widely used in electronic systems, but should *never* be used with fuel cells. The voltage is dropped by simply converting the surplus voltage into heat. Fuel cells will always be used where efficiency is of paramount importance, and linear regulators have no place in such systems.

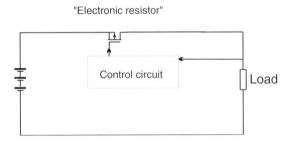

Figure 9.5 Linear regulator circuit

Because fuel cells are essentially low voltage devices, it is often desirable to step-up or boost the voltage. This can also be done quite simply and efficiently using switching circuits. The circuit of Figure 9.6 is the basis usually used.

We start our explanation by assuming some charge is in the capacitor. In Figure 9.6(a) the switch is on, and an electric current is building up in the inductor. The load is supplied by the capacitor discharging. The diode prevents the charge from the capacitor flowing back through the switch. In Figure 9.6(b) the switch is off. The inductor voltage rises sharply, because the current is falling. As soon as the voltage rises above that of the capacitor (plus about 0.6 volts for the diode) the current will flow through the diode, and charge up the capacitor and flow through the load. This will continue as long as there is still energy in the inductor. The switch is then closed again, as in Figure 9.6(a), and the inductor re-energised while the capacitor supplies the load.

Higher voltages are achieved by having the switch off for a short time. It can be shown that for an ideal convertor with no losses:-

$$V_2 = \frac{t_{ON} + t_{OFF}}{t_{OFF}} V_1 \qquad\qquad [9.2]$$

In practice the output voltage is somewhat less than this. As with the step-down (buck) switcher, control circuits for such boost or step-up switching regulators are readily available from many manufacturers.

The losses in this circuit come from the same sources as for the step-down regulator. However, because the currents through the inductor and switch are higher than the output current, the losses are higher. Also, all the charge passes through the diode this time, and so is subject to the 0.6 volts drop and hence energy loss. The result is that the efficiency of these boost regulators is somewhat less than for the buck. Nevertheless, over 80% should normally be obtained, and in systems where the initial voltage is higher (over 100 volts), efficiencies of 95% or more are possible.

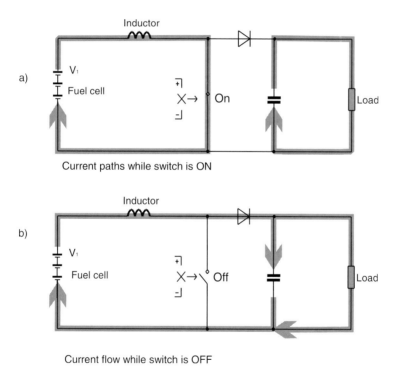

Figure 9.6 Circuit diagram to show the operation of a switch mode boost regulator.

A third possibility is to use a buck-boost regulator. In this case the final output is set somewhere within the operating range of the fuel cell. While such circuits are technically possible, their efficiency tends to be rather poor, certainly no better than the boost chopper, and often worse. The consequence is that this is not a good approach.

An exception to this is in cases where a small variation in output voltage can be tolerated, and an up-chopper circuit is used at *higher currents only*. This is illustrated in

Figure 9.7. At lower currents the voltage is not regulated. The circuit of Figure 9.6 is used, with the switch permanently off. However, the converter starts operating when the fuel cell voltage falls below a set value. Since the voltage shift is quite small, the efficiency would be higher.

It should be pointed out that, of course, the current out from a step-up converter is less than the current in. In Figure 9.7, if the fuel cell is operating at point A, the output will be at point A' - a higher voltage but a lower current. Also, the system is not entirely 'loss-free' while the converter is not working. The current would all flow through the inductor and the diode, resulting in some loss of energy.

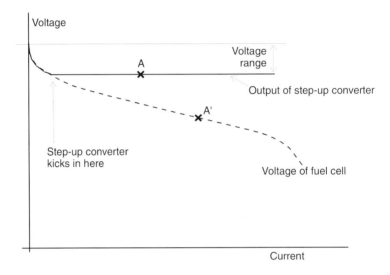

Figure 9.7 Graph of voltage against current for a fuel cell with a step-up chopper circuit that regulates to a voltage a little less than the maximum stack voltage.

These step-up and step-down switching or chopper circuits are called DC-DC converters. The symbol usually used in diagrams is shown in Figure 9.8. Complete units, ready made and ruggedly packaged, are available in a very wide range of powers and input and output voltages. In the cases where the requirements cannot be met by an off-the-shelf unit, we have seen that the units are essentially quite simple, and not hard to design. Standard control integrated circuits can nearly always be used to provide the switching signals.

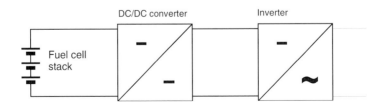

Figure 9.8 Typical fuel cell, DC-DC converter and inverter arrangement.

9.3   Inverters

9.3.1   Single Phase

Fuel cells will often be used in combined heat and power systems in both homes and businesses. In these systems the fuel cell will need to connect to the AC mains grid. The output may also be converted to AC in some grid independent systems.

In small domestic systems the electricity will be converted to a single AC voltage. In larger industrial systems the fuel cell will link to a 3-phase supply, as discussed in Section 9.3.2 below.

The arrangement of the key components of single phase inverter is shown in Figure 9.9. There are four electronic switches, labelled A, B, C and D, connected in what is called an 'H-bridge'. Across each switch is a diode, whose purpose will become clear later. The load through which the AC is to be driven is represented by a resistor and an inductor.

The basic operation of the inverter is quite simple. First switches A and D are turned on, and a current flows to the right through the load. These two switches are then turned off, - at this point we see the need for the diodes. The load will probably have some inductance, and so the current will not be able to stop immediately, but will continue to flow in the same direction, through the diodes across switches B and C, back into the supply. The switches B and C are then turned on, and a current flows in the opposite direction, to the left. When these switches turn off, the current 'free-wheels' on through the diodes in parallel with switches A and D.

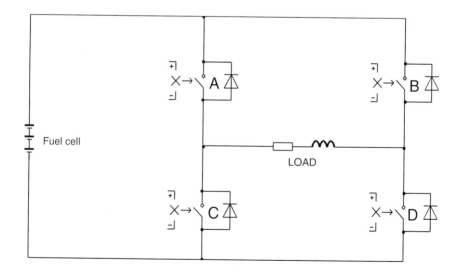

Figure 9.9 H-bridge inverter circuit for producing single phase alternating current.

The resulting current waveform is shown in Figure 9.10. In some cases, though increasingly few, this waveform will be adequate. The fact that it is very far from a sinewave will be a problem in almost all circumstances where there is a connection to the mains grid.

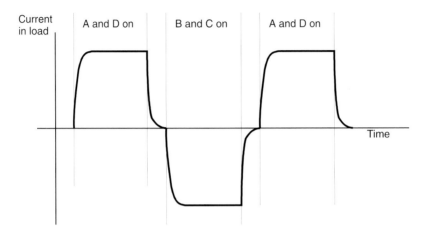

Figure 9.10 Current/time graph for a square wave switched single phase inverter.

The difference between a pure sinewave and any other waveform is expressed using the idea of 'harmonics'. These are sinusoidal oscillations of voltage or current whose frequency, f_v, is a whole number multiple of the fundamental oscillation frequency. It can be shown that *any* periodic waveform of *any* shape can be represented by the addition of harmonics to a fundamental sine wave. The process of finding these harmonics is known as Fourier analysis. For example, it can be shown that a square wave of frequency f can be expressed by the equation:

$$v = \sin(\omega t) - \frac{1}{3}\sin(3\omega t) + \frac{1}{5}\sin(5\omega t) - \frac{1}{7}\sin(7\omega t) + \frac{1}{9}\sin(9\omega t)....etc. \quad [9.3]$$

$$\text{where} \quad \omega = 2\pi f t$$

So, the difference between a voltage or current waveform and a pure sinewave may be expressed in terms of higher frequency harmonics imposed on the fundamental frequency. The problem is that these higher frequency harmonics can have harmful effects on other equipment connected to the grid, and on cables and switchgear due to high harmonic currents. Among the most serious are possible damage to protective equipment and disturbance of control systems. They can also cause inefficiencies in electric motors, unpleasant noise in all types of machine, and damage to computers and other electronic equipment, picture interference in TV sets and other undesirable effects. For this reason there are now regulations concerning the 'purity' of the waveform of an AC current supplied to the grid. Unfortunately, these standards vary in different countries and circumstances. However they are all expressed in terms of the amplitude of each harmonic relative to the amplitude of the fundamental frequency. One widely accepted standard is

IEC 1000-2-2 (Heier, 1998). The maximum percentage of each harmonic, up to the 50th, for this standard is given below in Table 9.2.

Table 9.2 Maximum permitted harmonic levels as a percentage of the fundamental in the low voltage grid up to the harmonic number 50 according to IEC 1000-2-2. (Reproduced from Heier, 1998)

ν		2	3	4	5	6	7	8	9	10
%		2.0	5.0	1.0	6.0	5.0	5.0	0.5	1.5	0.5
ν	11	12	13	14	15	16	17	18	19	20
%	3.5	0.2	3.0	0.2	0.3	0.2	2.0	0.2	1.5	0.2
ν	21	22	23	24	25	26	27	28	29	30
%	0.2	0.2	1.5	0.2	1.5	0.2	0.2	0.2	0.63	0.2
ν	31	32	33	34	35	36	37	38	39	40
%	0.6	0.2	0.2	0.2	0.55	0.2	0.53	0.2	0.2	0.2
ν	41	42	43	44	45	46	47	48	49	50
%	0.5	0.2	0.49	0.2	0.2	0.2	0.46	0.2	0.45	0.2

As can be seen from equation 9.3, a square wave exceeds the limits on the third harmonic by a factor of over 6, as well as others. So, how can a purer sinusoidal voltage and current waveform be generated? Two approaches are used, pulse width modulation and the more modern tolerance band pulse inverter technique

The principle of pulse width modulation is shown in Figure 9.11. The same circuit as shown in Figure 9.9 is used. In the positive cycle only switch D is on all the time, and switch A is on intermittently. When A is on, current builds up in the load. When A is off, the current continues to flow, because of the load inductance, through switch D and the "free-wheeling" diode in parallel with switch C, around the bottom right loop of the circuit.

In the negative cycle a similar process occurs, except that switch B is on all the time, and switch C is 'pulsed'. When C is on current builds in the load, when off it continues to flow – though declining – through the upper loop in the circuit, and through the diode in parallel with switch A.

The precise shape of the waveform will depend on the nature (resistance, inductance, capacitance) of the load, but a typical half cycle is shown in Figure 9.12. The waveform is still not a sinewave, but is a lot closer than that of Figure 9.10. Clearly, the more pulses there are in each cycle, the closer will be the wave to a pure sinewave, and the weaker will be the harmonics. Twelve pulses per cycle is a commonly used standard, and generally this gives satisfactory results.

In modern circuits the switching pulses are generated by microprocessor circuits. This has led to the adoption of a more 'intelligent' approach to the switching of inverters called the "tolerance band pulse" method. The method is illustrated in Figure 9.13. The output voltage is continuously monitored, and compared with an internal 'upper limit' and 'lower limit', which are sinusoidal functions of time. In the positive cycle, switch D (in Figure 9.9) is on all the time. Switch A is turned on, and the current through the load rises. When it reaches the upper limit A is turned off, and the current flows on, though declining, through the diode in parallel with C, as before. When the lower limit is reached, switch A is turned on again, and the current begins to build up again. This process is continuously repeated, with the voltage rising and falling between the tolerance bands.

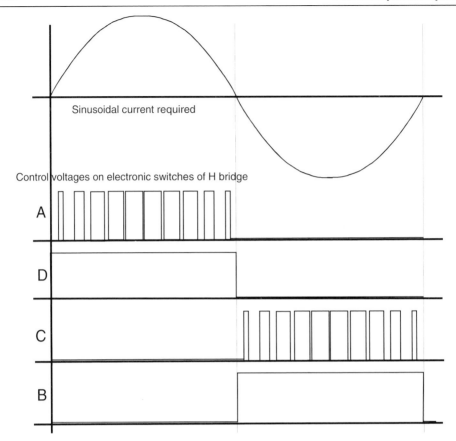

Figure 9.11 Pulse width modulation switching sequence for producing an approximately sinusoidal alternating current from the circuit of Figure 9.9

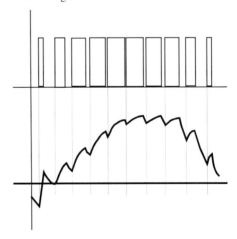

Figure 9.12 Typical voltage/time graph for a pulse modulated inverter

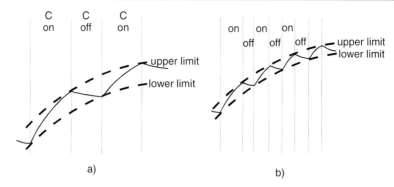

Figure 9.13   Typical voltage/time waveforms when using the tolerance band pulse inverter technique.

In Figure 9.13 the on/off cycle is shown in (a) for a wide tolerance band, and in (b) for a narrow tolerance. It should also be appreciated that the resistance and inductance of the load will also affect the waveform, and hence the frequency at which switching occurs. This is thus an adaptive system, that always keeps the deviation from a sinewave, and hence the unwanted harmonics, below fixed levels.

The only significant disadvantage of the "tolerance band" regulation method is that it is possible for the pulsing frequency to become very high. Because most of the losses in the system occur at the time of switching, while the transistors move from off to on and on to off, this can lead to lower efficiency. The well tried pulse width modulation method is still widely used, but the tolerance band technique is becoming more widespread, especially for mains connected inverters, where the nature of the load will not vary so much as in some other cases.

9.3.2   Three phase

In almost all parts of the world electricity is generated and distributed using three parallel circuits, the voltage in each one being out of phase with the next by 120°. While most homes are supplied with just one phase, most industrial establishments have all three phases available. So, for industrial combined heat and power systems the DC from the fuel cell will need to be converted to 3-phase AC.

This is only a little more complicated than single phase. The basic circuit is shown in Figure 9.14. Six switches, with free-wheeling diodes, are connected to the 3-phase transformer on the right. The way in which these switches are used to generate 3 similar but out of phase voltages is shown in Figure 9.15. Each cycle can be divided into six steps. The graphs of Figure 9.16 show how the current in each of the three phases changes with time using this simple arrangement. These curves are obviously far from sinewaves. In practice the very simple switching sequence of Figure 9.15 is modified using pulse-width modulation or tolerance band methods, in the same way as for the single phase inverters described above.

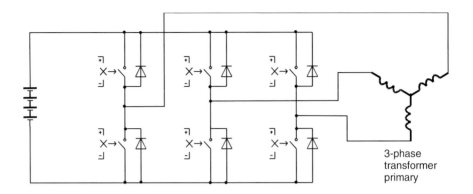

Figure 9.14   3-phase inverter circuit

Figure 9.15 (opposite) Switching pattern to generate 3-phase alternating current. ➡️

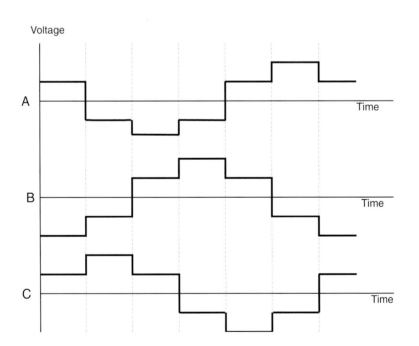

Figure 9.16 Current/time graphs for the simple 3-phase AC generation system shown in Figure 9.15, assuming a resistive load. One complete cycle for each phase is shown. Current flowing *out* from the common point is taken as *positive*.

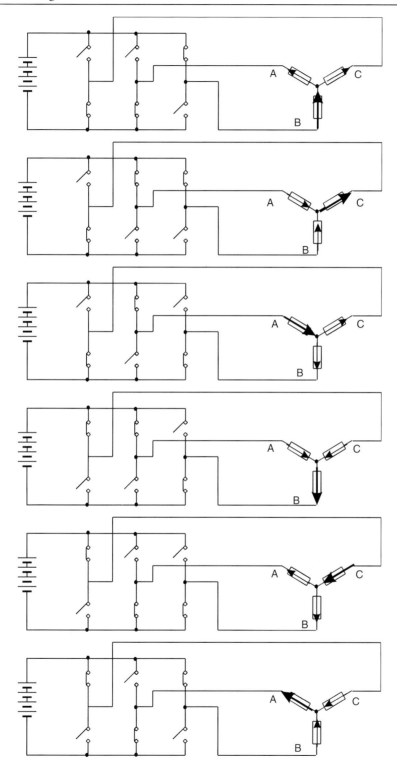

The modern three-phase inverter is made along similar lines whether it is for high or low power, and whether it is "line-commutated" (that is, the timing signals are derived from the grid to which it is connected) or "self-commutated" (i.e. independent of the grid). Indeed the same basic circuit is used whatever modulation method is used (pulse-width or tolerance band). That basic circuit is shown in Figure 9.17. The signals used to turn on and off the switches are taken from the microprocessor. Voltage and current sense signals may be taken from the 3 phases, the input, each switch, or other places. Digital signals from other sensors may be used. Also, instructions and information may be sent to and received from other parts of the system. The use of this information may be different in every case, but the *hardware* will be essentially the same – the circuit of Figure 9.17. Inverter units have thus become like many other electronic systems – a standard piece of hardware which is programmed for each application. The switches would vary depending on the power requirements, but the same controller could be suitable for a large range of powers. The same inverter hardware, differently programmed, would thus be suitable for fuel cells, solar panels, wind driven generators, Stirling cycle engines and any other distributed electricity generator. Inverters thus follow the trend seen throughout electronic engineering – lower prices, less electronics circuit design, more programming.

9.3.3   Regulatory issues and tariffs

It is not sufficient simply to generate AC electricity and connect your generator to the grid. Quite properly, any private system has to conform to certain standards before it can be allowed to supply electricity. One of the most important of these is the level of the harmonics, an issue that we have already discussed. However, there must also be protection systems to protect both the fuel cell, the inverter, and the grid from faults such as short circuits and lightning strikes. A particularly important feature is that the CHP system must disconnect from the grid in the event of the main power to the grid failing. This is to protect the grid service personnel when carrying out fault repair or maintenance, and to prevent the generator from feeding a short circuit with a damaging high current, the so-called ' fault current.' A generator is capable of feeding a local fault with considerably increased currents, when compared with that from a distant central power station. Switchgear and cabling may need expensive upgrading to cope with these increased currents that may flow under fault conditions. The local utility will have its own regulations and standards that equipment must conform to. In the UK one such is the G59 standard. In the USA equipment will nearly always need to be 'UL listed'. (UL refers to the Underwriters' Laboratory, a safety standards body.)

Where a fuel cell is generating for use in a mains electric grid, this will always be as part of a combined heat and power (CHP) system. The production of electricity will thus be linked with the production of heat. It is very likely that the demands for these two quantities will not be entirely synchronised, and so there will be times when the system may be producing less electricity than needed, or there may be times of high heat demand when there is an excess of electricity supply. Such problems can be solved by maintaining a connection to the local mains utility grid, and taking and supplying electricity as needed. The vast majority of CHP systems do this.

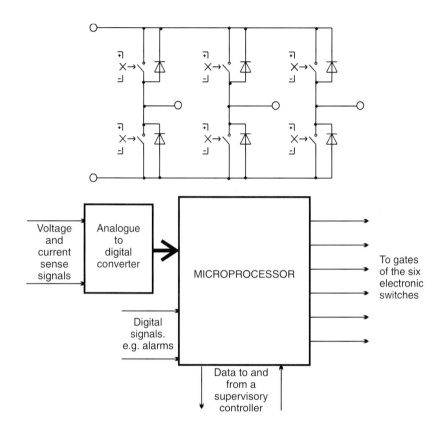

Figure 9.17 "Universal" 3-phase inverter circuit.

The precise regulations covering grid connection (intertie) equipment will, unfortunately, vary between countries, and between states within countries. In the USA the Public Utilities Regulatory Policies Act (PURPA) of 1978 requires electric utilities to purchase electricity generated by small-power producers (SPPs) who run CHP systems. Furthermore, the price of this electricity is fixed by a body independent of the electric utilities, a state utility regulatory agency, which operates under federal guidelines. The concept of "avoided cost" is used as the basis for the utility power purchase rates. This is the rate that it costs the utility to generate the electricity itself. However, those rates will naturally be lower than the 'retail' rate at which they sell power.

For this reason a CHP system connected to the grid will usually have two meters – one measuring electricity consumed, the other electricity provided. The charges paid will sometimes vary according to time. It is generally more expensive to generate and supply electricity during periods of high demand. Peaks of demand occur daily and vary according to the types of client a utility has, on climate, and on weather conditions. CHP systems are particularly suited to helping with the winter mornings peak that occurs in many areas, though they are not so well suited to reduce the summer afternoon peak that can occur in warmer climes. Such time of use (TOU) charges may apply to both electricity bought and sold, which could be advantageous to some CHP systems.

The two meter system puts the SPP at a disadvantage[1]. However, in order to reduce investment in power generation equipment, and to encourage the development of SPPs with their environmental advantages, some utilities and states in the USA have encouraged and instituted so-called "net-metering". One meter is used, which winds back when electricity is supplied to the grid. The SPP is thus effectively paid the same for the electricity it supplies as it pays for what it consumes. Political pressure for the extension of these more favourable billing arrangements is likely in the years ahead, and is certainly well under way in the USA. However, this should not be overstated, as many US utilities are only required to purchase a very small percentage of their electricity via net metering, typically 0.1%.

9.3.4   Power Factor Correction

A major advantage of distributed generators of electricity, especially those of higher power, is the ability to correct the power factor of a customer, or even a small district. The problem of power factors arises when the current consumed is not exactly in phase with the voltage. For example the current might lag behind the voltage, as in Figure 9.18. The current can be split into two components, an in-phase component and a 90° out of phase component. This 90° out of phase component consumes no power, and is known as 'reactive' power. Although it consumes no power locally, it does increase the power dissipated in the resistive electric cables distributing the electricity. This reactive power is, in a sense, 'free' to *generate* but not free to *distribute*. For this reason it is particularly advantageous if it can be generated locally. Inverters fed from DC supplies are particularly suitable for this. The switching pulses in the PWM or tolerance band inverter circuits are driven so that the current produced is deliberately out of phase with the line voltage.

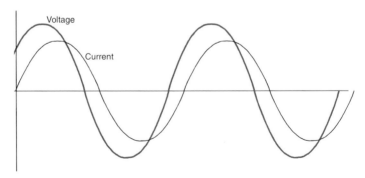

Figure 9.18 Voltage and current out of phase. The reactive power can be locally generated by distributed power systems such as fuel cells.

This reactive power can be generated by other small electricity generators, such as wind driven systems. However, it is a particular feature of fuel cell CHP systems, as they will always be very close to the point of usage of electricity.

---

[1] For example, in Oxford, in 2000, the local utility charges 6p and pays only 2p, per kWh.

9.4   Electric Motors

9.4.1   General points

The importance of electric motors in fuel cell systems can hardly be overstated. For example, the 2kW system described in Section 4.7, and shown in Figure 4.29, features three electric motors just to make it work. The final electrical power will often be delivered to an electric motor also. This is certainly the case for the 200 kW system also outlined in Section 4.7, which features at least four electric motors.

Electric motors are of course very widely used in engineering products. For example, there are about 20 electric motors in a typical modern motor car. Most of these motors are ordinary "brushed" DC motors, using permanent magnets. It is assumed that the reader is familiar with this type of motor. They are very suitable for occasional use applications – which is the common mode of use in, for example, motor vehicles. To take a specific case, what is the lifetime use of the motors used to adjust the position of the external mirrors of a car? A few minutes at most. If motors of this type have a short life, or are inefficient, it is not of great importance. The motors used in fuel cells are *not* like this, they typically circulate reactant gases or cooling fluids, and are in use *all the time* the fuel cell is in use, and so they should be of the highest possible efficiency, and have the longest possible life. In addition, the presence of volatile fuels like hydrogen means that the sparks that inevitably arise occasionally with brushes should be avoided. For these reasons we will only be considering "brushless" motors in this section.

At least half the electricity generated in the power systems of developed countries is consumed by just one type of electric motor - the induction motor. (Walters, 1999) This type of motor requires an AC supply, three phase for preference. Such a three phase supply can easily be generated by an inverter as described above in Section 9.3.2, and so they are sometimes used with fuel cells. A brief description of this type of motor is given in Section 9.4.2. Although hugely successful, and generally highly efficient, the induction motor is not the best in terms of power density and efficiency. Two other modern motors are also used in fuel cell systems; these are the brushless DC motor described in Section 9.4.3, and the switched reluctance motor described in Section 9.4.4.

9.4.2   The induction motor

The induction motor is very widely used in industrial machines of all types. Its technology is very mature. Induction motors require an AC supply, which might make them seem unsuitable for a DC source such as fuel cells. However, as we have seen, AC can easily be generated using an inverter, and in fact the inverter needed to produce the AC for an induction motor is no more complicated or expensive than the circuits needed to drive the brushless DC or switched reluctance motors. So, these widely available and very reliable motors are suited to use with fuel cells.

The principle of operation of the three-phase induction motor is shown in Figures 9.19 and 9.20. Three coils are wound right around the outer part of the motor, knows as the stator, as shown in the top of Figure 9.19. The rotor usually consists of copper or aluminium rods, all electrically linked (short circuited) at the end, forming a kind of cage, as also shown in Figure 9.19. Although shown hollow, the interior of this cage rotor will usually be filled with laminated iron.

The three windings are arranged so that a positive current produces a magnetic field in the direction shown in Figure 9.20. If these three coils are fed with a 3 phase alternating current, as in Figure 9.20, the resultant magnetic field rotates anti-clockwise, as shown at the bottom of Figure 9.20.

This rotating field passes through the conductors on the rotor, generating an electric current.

A force is produced on these conductors carrying an electric current, which turns the rotor. It tends to 'chase' the rotating magnetic field. If the rotor were to go at the same speed as the magnetic field, there would be no relative velocity between the rotating field and the conductors, and so no induced current, and no torque. The result is that the torque speed graph for an induction motor has the characteristic shape shown in Figure 9.21. The torque rises as the angular speed 'slips' behind that of the magnetic field, up to an optimum slip, after which the torque declines somewhat.

The winding arrangement of Figure 9.19 and 9.20 is knows as "2-pole". It is possible to wind the coils so that the magnetic field has 4, 6, 8 or any even number of poles. The speed of rotation of the magnetic field is the supply frequency divided by the number of pole pairs. So, a 4-pole motor will turn at half the speed of a 2-pole motor, given the same frequency AC supply, a 6-pole motor a third the speed, and so on. This gives a rather inflexible way of controlling speed. A much better way is to control the frequency of the 3-phase supply. Using a circuit such as that of Figure 9.17 this is easily done. The frequency does not precisely control the speed, as there is a 'slip' depending on the torque. However, if the angular speed is measured, and incorporated into a feedback loop, the frequency can be adjusted to attain the desired speed.

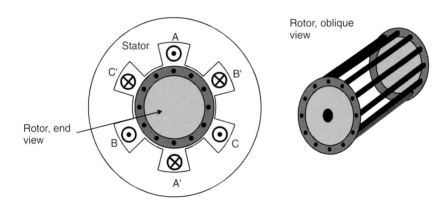

Figure 9.19 Diagram showing the stator and rotor of an induction motor

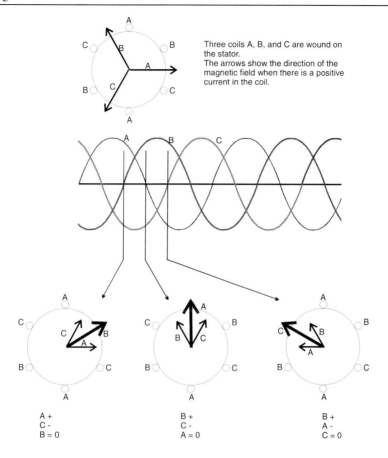

Figure 9.20 Diagrams to show how a rotating magnetic field is produced within an induction motor.

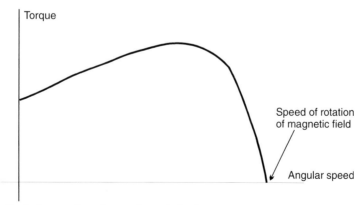

Figure 9.21 Typical torque/speed curve for an induction motor.

The maximum torque depends on the strength of the magnetic field in the gap between the rotor and the coils on the stator. This depends on the current in the coils. A problem is that as the frequency increases the current reduces, if the voltage is constant, because of the inductance of the coils having an impedance that is proportional to the frequency. The result is that, if the inverter is fed from a fixed voltage, the maximum torque is inversely proportional to the speed. This is liable to be the case with a fuel cell system.

Induction motors are very widely used. Very high volume of production makes for a very reasonably priced product. Much research has gone into developing the best possible materials. Induction motors are as reliable and well developed as any technology. However, the fact that a current has to be induced in the motor adds to the losses, with the result that induction motors tend to be a little (1 or 2%) less efficient than the motors to be described below.

9.4.3   The brushless DC motor

The brushless DC motor (BLDC motor) is really an AC motor! The current through it alternates, as we shall see. It is called a "brushless DC motor" because the alternating current *must* be variable frequency and so derived from a DC supply, and because its speed/torque characteristics are very similar to the ordinary "with brushes" DC motor. As a result of "brushless DC" being not an entirely satisfactory name, it is also, very confusingly, given different names by different manufacturers and users. The most common of these is "self-synchronous AC motor", but others include "variable frequency synchronous motor", "permanent magnet synchronous motor", and "electronically commutated motor" (ECM).

The basis of operation of the BLDC motor is shown in Figure 9.22. Switches direct the direct from a DC source through a coil on the stator. The rotor consists of a permanent magnet. In Figure 9.22(a) the current flows in the direction that magnetises the stator so that the rotor is turned clockwise, as shown. In 9.22(b) the rotor passes between the poles of the stator, and the stator current is switched off. Momentum carries the rotor on, and in 9.22(c) the stator coil is re-energised, but the current and hence the magnetic field, are reversed. So the rotor is pulled on round in a clockwise direction. The process continues, with the current in the stator coil alternating.

Obviously, the switching of the current must be synchronised with the position of the rotor. This is done using sensors. These are often Hall effect sensors that use the magnetism of the rotor to sense its position, but optical sensors are also used.

A problem with the simple single coil system of Figure 9.22 is that the torque is very unsteady. This is improved by having three (or more) coils, as in Figure 9.23. In this diagram coil B is energised to turn the motor clockwise. Once the rotor is between the poles of coil B, coil C will be energised, and so on.

The electronic circuit used to drive and control the coil currents is usually called an inverter – and it will be the same as, or very similar to, our "universal inverter" circuit of Figure 9.17. The main control inputs to the microprocessor will be the position sense signals.

A feature of these BLDC motors is that the torque will reduce as the speed increases. The rotating magnet will generate a back e.m.f. in the coil which it is approaching. This back e.m.f. will be proportional to the speed of rotation, and will reduce the current flowing in the coil. The reduced current will reduce the magnetic field strength, and hence the torque. Eventually the size of the induced back e.m.f. will equal the supply voltage, and at this point the maximum speed has been reached. Notice also that this type of motor can very simply be used as a generator of electricity, and for regenerative or dynamic braking.

Although the current through the motor coils alternates, there must be a DC supply, which is why these motors are generally classified as 'DC'. They are very widely used in computer equipment to drive the moving parts of disc storage systems and fans. In these small motors the switching circuit is incorporated into the motor with the sensor switches. However, they are also used in higher power applications, with more sophisticated controllers (as of Figure 9.17), which can vary the coil current (and hence torque) and thus produce a very flexible drive system. Some of the most sophisticated electric vehicle drive motors are of this type, and one is shown in Figure 9.24. This is a 100 kW, oil cooled motor, weighing just 21kg.

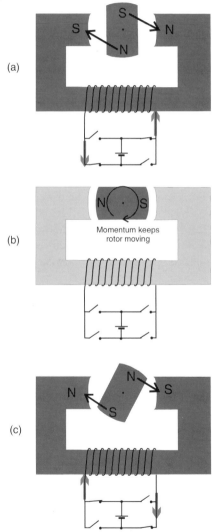

Figure 9.22 Diagram showing the basis of operation of the brushless DC motor.

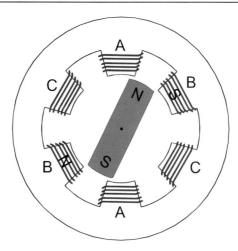

Figure 9.23 Diagram showing an arrangement of three coils on the stator of a BLDC motor.

These BLDC motors need a strong permanent magnet for the rotor. The advantage of this is that currents do not need to be induced in the motor (as with the induction motor), making them somewhat more efficient and giving a slightly greater specific power. The control electronics needed is essentially identical as for the induction motor. However, the permanent magnet rotor does add significantly to the cost of these motors.

Figure 9.24 100 kW, oil cooled BLDC motor for automotive application. This unit weighs just 21 kg. (Photograph reproduced by kind permission of Zytek Ltd.)

9.4.4 Switched reluctance motors

Although only recently coming into widespread use, the switched reluctance (SR) motor is, in principle, quite simple. The basic operation is shown in Figure 9.25 below. In 9.25(a) the iron stator and rotor are magnetised by a current through the coil on the stator. Because the rotor is out of line with the magnetic field a torque will be produced to minimise the air gap and make the magnetic field symmetrical. We could lapse into rather 'medieval' science and say that the magnetic field is 'reluctant' to cross the air gap, and seeks to minimise it. Medieval or not, this is why this type of motor is called a reluctance motor.

At the point shown in Figure 9.25(b) the rotor is aligned with the stator, and the current is switched off. Its momentum then carries the rotor on round over ¼ of a turn, to the position of 9.25(c). Here the magnetic field is re-applied, in the same direction as before. Again, the field exerts a torque to reduce the air gap and make the field symmetrical, which pulls the rotor on round. When the rotor lines up with the stator again, the current would be switched off.

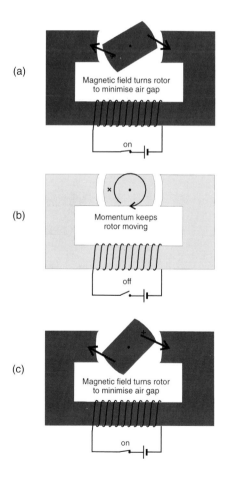

Figure 9.25 Diagram showing the principle of operation of the switched reluctance motor.

In the switched reluctance motor, the rotor is simply a piece of magnetically soft iron. Also, the current in the coil does not need to alternate. Essentially then, this is a very simple and potentially low cost motor. The speed can be controlled by altering the length of time that the current is on for in each "power pulse". Also, since the rotor is not a permanent magnet, there is no back e.m.f. generated in the way it is with the BLDC motor, which means that higher speeds are possible. In the fuel cell context, this makes the SR motor particularly suitable for radial compressors and blowers.

The main difficulty with SR motor is that the timing of the turning on and off of the stator currents must be much more carefully controlled. For example, if the rotor is $90°$ out of line, as in Figure 9.22(a), and the coil is magnetised, no torque will be produced, as the field would be symmetrical. So, the torque is much more variable, and as a result early SR motors had a reputation for being noisy.

The torque can be made much smoother by adding more coils to the stator. The rotor is again laminated iron, but has 'salient poles' i.e. protruding lumps. The number of salient poles will often be two less than the number of coils. Figure 9.26 shows the principle. In 9.26(a) coil A is magnetised, exerting a clockwise force on the rotor. When the salient poles are coming into line with coil A, the current in A is switched off. Two other salient poles are now nearly in line with coil C, which is energised, keeping the rotor smoothly turning. Correct turning on and off of the currents in each coil clearly needs good information about the position of the rotor. This is usually provided by sensors, but modern control systems can do without these. The position of the rotor is inferred from the voltage and current patterns in the coils. This clearly requires some very rapid and complex analysis of the voltage and current waveforms, and is achieved using a special type of microprocessor called a digital signal processor[2].

An example of a rotor and stator from an SR motor is shown in Figure 9.27. In this example the rotor has eight salient poles.

The stator of an SR motor is similar to that in both the induction and BLDC motor. The control electronics is also similar – a microprocessor and some electronic switches, along the lines of Figure 9.17. However, the rotor is significantly simpler, and so cheaper and more rugged. Also, when using a core of high magnetic permeability the torque that can be produced within a given volume exceeds that produced in induction motors (magnetic action on current) and BLDC motors (magnetic action on permanent magnets) (Kenjo, 1991, p.161). Combining this with the possibilities of higher speed means that a higher power density is possible. The greater control precision needed for the currents in the coils makes these motors somewhat harder to apply on a "few-of" basis, with the result that they are most widely used in cost-sensitive mass-produced goods such as washing machines and food processors. However, we can be sure that their use will become much more widespread.

Although the peak efficiency of the SR motor may be slightly below that of the BLDC motor, SR motors maintain their efficiency over a wider range of speed and torque than any other motor type.

---

[2] Although they were originally conceived as devices for processing audio and picture signals, the control of motors is now a major application of digital signal processors. BLDC motors can also operate without rotor position sensors in a similar way.

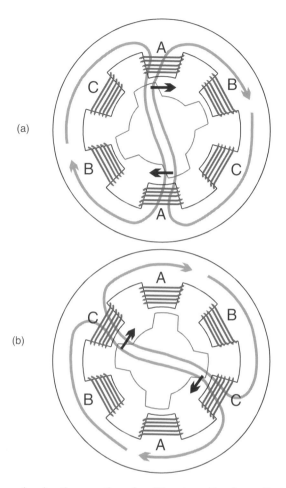

(a)

(b)

Figure 9.26 Diagram showing the operation of an SR motor with a four salient pole rotor.

Figure 9.27 The rotor and stator from an SR motor. (Photograph reproduced by kind permission of SR Drives Ltd.)

9.4.5   Motor efficiency

It is clear that the motor chosen for any application should be as efficient as possible. How can we predict what the efficiency of a motor might be? It might be supposed that the *type* of motor chosen would be a major factor, but in fact it is not. Other factors are much more influential than whether the motor is BLDC, SR or induction.

An electric motor is, in energy terms, fairly simple. Electrical power is the input, and mechanical work is the desired output, with some of the energy being converted into heat. The input and output powers are straightforward to measure - the product of voltage and current for the input, and torque and angular speed at the output. However, the efficiency of an electric motor is not so simple to measure and describe as might be supposed. The problem is that it can change markedly with different conditions, and there is no single internationally agreed method of stating the efficiency of a motor[3]. (Auinger, 1999) Nevertheless it is possible to state some general points about the efficiency of electric motors - the advantages and disadvantages of the different types, and the effect of motor size.

The first general point is that motors become more efficient as their *size* increases. Table 9.3 below shows the efficiency of a range of 3-phase, 4-pole induction motors. The efficiencies given are the minimum to be attained before the motor can be classified "Class 1" efficiency under European Union regulations. The figures clearly show the effect of size. While these figures are for induction motors, exactly the same effect can be seen with other motor types, including BLDC and SR.

Table 9.3 The minimum efficiency of 4-pole 3-phase induction motors to be classified as Class 1 efficiency under EU regulations. Efficiency measured according to IEC 34.2.

Power kW	Minimum efficiency %
1.1	83.8
2.2	86.4
4	88.3
7.5	90.1
15	91.8
30	93.2
55	94.2
90	95.0

The second factor that has more control over efficiency than motor type is the *speed* of a motor. Higher speed motors are more efficient than lower. The reason for this is that one of the most important losses in a motor is proportional to torque, rather than power, and a lower speed motor will have a higher torque, for the same power, and hence higher losses.

A third important factor is the cooling method. Motors that are liquid cooled run at lower temperatures, which reduces the resistance of the windings, and hence improves efficiency, though this will only affect things by about 1%.

---

[3] The nearest to such a standard is IEC 34-2.

Another important consideration is that the efficiency of an electric motor might well be very different from any figure given in the specification, if it operates well away from optimum speeds and torque. In some cases an efficiency map, like that of Figure 9.28 may be provided. This is based on a real BLDC motor. The maximum efficiency is 94%, but this efficiency is only obtained for a fairly narrow range of conditions. It is perfectly possible for the motor to operate at well below 90% efficiency.

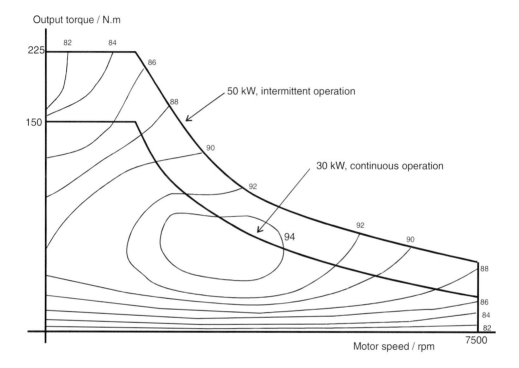

Figure 9.28 The efficiency map for a 30 kW BLDC motor.

So, we can say that the efficiency of a good quality motor will be quite close to the figures given in Table 9.3 for all motor types, even if they are not induction motors. The efficiency of the BLDC and SR motors is likely to be one or two percent higher than for an induction motor, since there is far less loss in the rotor. The SR motor manufacturers also claim that their efficiency is maintained over a wider range of speed and torque conditions.

9.4.6   Motor mass

A motor should generally be as small and light as possible while delivering the required power. As with the case of motor efficiency, the type of motor chosen is much less important than other factors when it comes to the specific power and power density of an electric motor.

Figure 9.29 below is a chart showing typical specific powers for different types of motor at different powers. Taking the example of the BLDC motor, it can be seen that the cooling method used is a very important factor. The difference between the air cooled and liquid cooled BLDC motor is most marked. The reason for this is that the size of the motor has to be large enough to dispose of the heat losses. If the motor is liquid cooled, then the same heat losses can be removed from a smaller motor.

We would then expect that efficiency should be an important factor. A more efficient motor could be smaller, since less heat disposal would be needed. This is indeed the case, and as a result all the factors that produce higher efficiency, and which were discussed in the previous section, also lead to greater specific power. The most important of these are:-

- Higher *power* leads to higher efficiency, and hence higher specific power. This can be very clearly seen in Figure 9.29. (However, note that the logarithmic scale tends to make this effect appear less marked.)

- Higher *speed* leads to higher power density. The size of the motor is most strongly influenced by the motor *torque*, rather than *power*. The consequence is that a higher speed, lower torque motor will be smaller. So if a low speed rotation is needed, a high speed motor with a gearbox will be lighter and smaller than a low speed motor. A good example is an electric vehicle, where it would be possible to use a motor directly coupled to the axle. However, this is not done, and a higher speed motor is connected by (typically) a 10:1 gearbox. Table 9.4 below shows this, by giving the mass of a sample of induction motors of the same power but different speeds.

- The more efficient *motor types*, SR and BLDC, have higher power density that the induction motor.

The curves of Figure 9.29 give a good idea of the likely power density that can be expected from a motor, and can be used to estimate the mass. The lines are necessarily broad, as the mass of a motor will depend on many factors other than those we have already discussed. The material the frame is made from is of course very important, as is the frame structure.

Table 9.4 The mass of some 37 kW induction motors, from the same manufacturer, for different speeds. The speed is for a 50 Hz AC supply

Speed (rpm)	Mass (kg)
3000	270
1500	310
1000	415
750	570

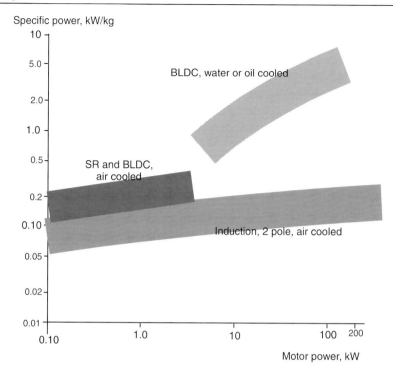

Figure 9.29 Chart to show the specific power of different types of electric motors at different powers. The power here is the continuous power. Peak specific powers will be about 50% higher. Note the logarithmic scales. (This chart was made using data from several motor manufacturers.)

9.5   Fuel Cell/Battery Hybrid Systems

The use of batteries in association with a fuel cell can reduce the cost of a fuel cell based power system. This is especially the case when powering certain types of electronic equipment.

The essence of a fuel cell/battery hybrid is that the fuel cell works quite close to its maximum power at all times. When the total system power requirements are low, then the surplus electrical energy is stored in a rechargeable battery. When the power requirements exceed those that can be provided by the fuel cell, then energy is taken from the battery. This presupposes that the power requirements are quite variable. Indeed, in cases where the power requirement is quite constant, then the fuel cell/battery hybrid system offers no serious advantages[4].

The easiest hybrid systems to design are those where the electrical power requirements are highly variable yet also predictable. Such a situation in illustrated in Figure 9.30(a), and can occur with certain data logging equipment, with data transmitters, certain types of telecommunications equipment, and with land or buoy based navigation equipment. For fairly long periods the device is in "standby" mode, and the fuel cell will be recharging the

---

[4] However, other hybrid systems, such as solar panel + fuel cell + battery might on occasion be attractive in such cases.

battery. During the "transmit" periods, most of the power is supplied by the battery. The fuel cell operates more or less continuously at the average power, and is thus simple to specify. The battery requirements are also clear; they must provide sufficient power, and hold enough energy for just one "high power pulse", and must be able to be recharged in the period between high power pulses.

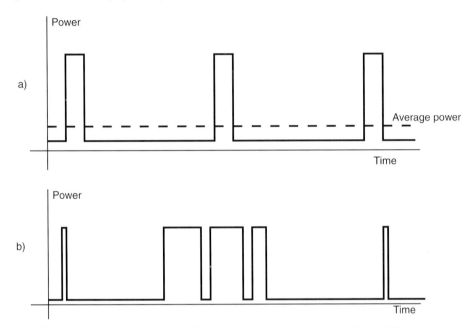

Figure 9.30 Power/time graphs for two different systems suited to a hybrid fuel cell/battery power supply.

In many more cases the electrical power requirements are not only highly variable but also unpredictable. A good example is the mobile telephone. Such a situation is shown in Figure 9.30(b) above. The higher power pulses can be more frequent and longer lasting, or less frequent and shorter. It would still be possible to have a fuel cell running at the long-term average power, but the battery would need to be considerably larger, since it might need to provide several high power pulses without any time to recharge. It might also be advantageous to increase the power of the fuel cell above the long term average power, so that the battery recharged more quickly.

In such hybrid systems the two main variables are thus the fuel cell power and the battery capacity. In the case of a mobile telephone these would be chosen on the basis of the following:-

- The standby power
- The "on-call" power
- The probability of receiving a call
- The probable length of a call
- The acceptable probability of system failure

Notice that the only way to bring this last factor to zero is to have the fuel cell power equal to the "on call" power, and abandon the hybrid concept altogether. Most likely, the system would be modelled on a computer.

In essence, such systems could be said to be using the fuel cell as a battery charger. This is liable to be a common early use of small fuel cells, either hydrogen power or methanol. The concept is shown in Figure 9.31. A controller of some sort in needed, to prevent overcharging. Also, as discussed in Section 9.2, a DC/DC converter would normally be necessary. Such a system would be particularly attractive as a range extender for mobile phones, and would suit the direct methanol fuel cell, since the average power, and hence fuel cell power, is so low. It could be very easy to refuel such a small fuel cell with methanol. The battery could be the standard mobile phone battery, with the fuel cell and its integral DC/DC converter in a separate charger unit (Hockaday, 1999). Portable computers are also another very likely candidate for such fuel cell/battery systems.

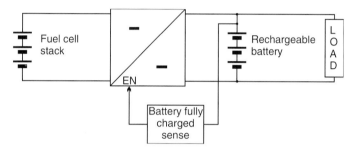

Figure 9.31 Simple fuel cell/battery hybrid system.

In some cases the models for designing such hybrid systems are quite well developed, because they have already been used for solar panel/battery hybrid systems. These are very widely used in telecommunications and navigation equipment. In comparison, the model needed for a fuel cell based system is much simpler, since with the solar panel both the consumption *and generation* of power vary in a mixture of random and predictable ways. With fuel cell based hybrid systems, the primary generation of electrical power can at least be relied upon and accurately modelled.

Such hybrid systems, where the peak or battery power is much greater than the average power (and hence fuel cell power) are sometimes known as "hard" hybrid systems. By contrast, a "soft" hybrid is one where the battery power and energy storage are quite low compared to the fuel cell power. Such systems might be found in vehicles and boats for example, where the peak power and the average power are not so different. An example power/time graph is shown in Figure 9.32(a) overleaf. The fuel cell power is sufficient most of the time, but the battery can "lop the peaks" off the power requirement, and can substantially reduce the required fuel cell capacity. During times when the fuel cell power exceeds the power demand, the battery is recharged, as before. This sort of power demand is characteristic of urban electric vehicles, where the peaks correspond to occasions such as accelerating from traffic lights, yet most of the time the vehicle will be proceeding slowly and steadily, or else stationary. A further possibility is illustrated in the power/time graph of Figure 9.32(b). Here an electric vehicle is using the motor as a brake – all the electric motors described in the previous section can be used in this way. In *dynamic* braking, the motor is used as a generator, the motion energy is converted into electrical energy, but is then simply passed through a resistor and converted to heat. In *regenerative* braking the electrical energy is passed to a rechargeable battery, to be used later to run the motor. Regenerative braking is clearly the better option from the point of view of system efficiency, but it does presuppose a hybrid system, with a rechargeable battery. Such a hybrid system would also need a fairly sophisticated control system. For one thing, the flow of power from the fuel cell to the battery would need to be properly controlled, since the situation is much more variable, with sometimes only quite a small amount of surplus fuel cell power. Secondly, the battery's "normal" state would have to be at somewhat less than fully charged, so that it could absorb the electrical energy supplied by the motors during braking. The motor controller would also need to be a full "four quadrant" type[5]. The controller would need to be more complex than the system of Figure 9.31. Figure 9.33 opposite shows in system diagram form what would be required. In particular, a measurement of the battery state of charge would be needed – rather than just a 'fully charged' indication. The energy flows to and from the different parts of the system are much more complex.

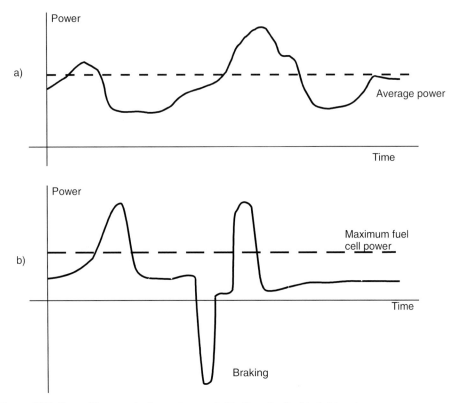

Figure 9.32 Power/time graphs for systems suitable for a "softer" hybrid system.

---

[5] Four quadrants refers to the four possible modes of motoring - forwards accelerating, braking while going forwards, backwards accelerating, braking while going backwards.

In this last case, where the battery is supplying fairly short term power peaks, and also absorbing power from regenerative braking, the rate of transfer of charge in and out of the battery is liable to be a problem. A way round this is to use special high capacity or "super" capacitors. These are being developed specially for such "soft" hybrid applications, both for fuel cells and internal combustion engines. Their energy density is much less than rechargeable batteries, but their power density is much greater, typically 2.5 kW.kg^{-1}, and they can also be used for at least 500,000 charge/discharge cycles (Harri et al., 1999).

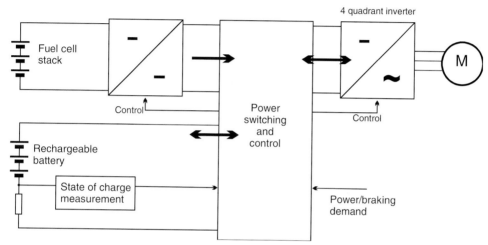

Figure 9.33 System diagram for a hybrid fuel cell/battery system. The bold arrows indicate energy flows. The resistor in series with the rechargeable battery is a current sense resistor used to measure the charge in and out.

The decision of whether or not to go for a hybrid system is influenced by two opposing characteristics of fuel cells. The main driving force in favour is that fuel cells are very expensive "per Watt", and so it makes sense to have them working at their full rated power for as much time as possible, to get the maximum possible value from the investment. On the other hand, one of the advantages of fuel cells over most other fuel energy converters is that at part load their voltage, and thus their efficiency[6], increases. Thus we should rate the fuel cell at the maximum required power, and if it operates at only part load for much of the time, so much the better: the efficiency will have improved. It is fair to say, that at the time of writing, the question of cost weighs more heavily in the balance.

In this discussion we have just considered fuel cells in partnership with an energy storage device. However, there are of course a huge number of other possibilities, for example, a solar panel, fuel cell and battery. Such systems have been successfully deployed in roadside variable message systems (Gibbard, 1999), and could have many other applications in remote power supplies. The fuel cell could compensate for the somewhat unreliable nature of solar energy in many situations.

---

[6] See Section 2.4

9.6   Example System

We close this chapter by presenting, in outline form, an example fuel cell system design that has been used successfully. It is the fuel cell engine for buses designed and manufactured by Ballard Power Systems. We have already mentioned this system in Chapter 4, and it is shown in Figure 4.30. Essentially the same system is shown below in Figure 9.34.

Some of the data in the account below is taken from product information provided by Ballard. *Such given data is printed in italics.* Other discussion is speculative, based on the theory outlined in this book.

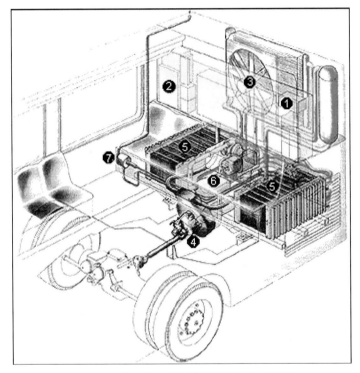

Figure 9.34 Fuel cell engine for buses based on 260 kW PEM fuel cell. (Diagram reproduced by kind permission of Ballard Power Systems.)

The operating voltage of the stack (5) is from about *450 to 750 volts. This is maintained to between 650 and 750 volts using an up-chopper* (boost switch mode regulator) as in Figure 9.6. This unit is labelled (1) in Figure 9.34 above. We can therefore suppose that the system being used is as outlined at the end of Section 9.2, and illustrated in Figure 9.7, with the converter kicking in when the voltage reaches 650 volts. There will also be several step-down regulators (as in Figure 9.3) to provide lower voltages for various subsystems such as the controller (2), and the cooling system (3). There is also a *12 Volt battery used when starting the system,* which will be charged from the fuel cell. We might estimate an overall efficiency of about 95% for these electrical sub-systems, making loses of about 13 kW.

The system is controlled using a programmable logic controller (PLC) (2). It is beyond the scope of this book to give an account of these very widely use control devices. They are used in all sorts of industrial control applications, and are programmed in "ladder logic" - an easy to learn symbolic programming method. One PLC can simultaneously control several control loops.

The fuel cell system is *water cooled, with a 'radiator' and electrically operated fan* (3). *The heat output is rated as 1.3 MBTU.hr⁻¹*, which is equivalent to 381 kW. This cooling system removes heat from the electric motor, and all the associated inverters and regulators, which are also water cooled. As we shall see, the reactant air supply to the fuel cell also has to be cooled. *An ion exchange filter is used to keep the cooling water pure, preventing it from becoming a conductor of electricity.* Clearly then, no anti-freeze can be used, so the *system must be kept above zero, which is done using a heater, connected to the mains when the bus is garaged.* We note that almost all the loses, including those in the air intake system and the electrical systems are dealt with by this cooling system. 381 kW therefore seems a reasonable waste heat rate for 260 kW fuel cell.

In Section 8.11 we introduced the idea of the 'effectiveness' of a cooling system. It was noted that a figure of between 20 and 30 should be obtainable. So, we can speculate that the cooling system, which disposes of up to 381 kW of heat might consume about 381/20, that is about 20 kW of electrical power. This would be used to drive the fan and the various pumps moving the water around the different sub-systems.

The electric motor (4) is a BLDC motor (Section 9.4.3), water cooled, rated at 160 kW continuous. The inverter circuit uses IGBTs. (See Section 9.2.1.) The inverter is also water cooled. The combined efficiency of the inverter and motor is 92%. 160 kW is the continuous power rating, so for short periods it will be able to produce higher powers. Exactly what the peak power is we cannot tell, but it will be in the region of 200 kW. There is evidence (Spiegel et al., 1999) that some models of this fuel cell engine for buses use water cooled induction motors, instead of BLDC. This illustrates very well what was said in the previous section – that there is not a great deal to choose between the three main brushless motor types. Induction motors are more rugged and lower cost, BLDC are slightly more efficient and compact. *Dynamic braking (see Section 9.5) is used to reduce wear on the vehicle brakes*, but this is not a hybrid system. *The electrical energy heats a resistor that is in the cooling system, and effectively works as an immersion heater.* This same resistor, as has been mentioned above, is used in very cold conditions to prevent the fuel cell freezing by being plugged into an external supply. *The motor is coupled to the forward running drive shaft via a 2.43:1 gear.* This is connected to the bus back axle via a differential, which will have a gear ratio of about 5:1.

The fuel cell stacks (5) are arranged as two units connected in parallel. Each half consists of 10 stacks, each of power 13 kW. The maximum electrical power is thus 260 kW. The cells are operated at 90 °C, and both the fuel and the air are humidified, as explained in Section 4.4.4 of Chapter 4. *The operating pressure of the stacks is increased when higher power is needed, up to a maximum of 207 kPa above ambient pressure.* Each half unit probably consists of about 750 cells in series. They are constructed in the mainstream fashion for PEMFC, that is with graphite, or graphite/polymer mixture, bipolar plates. They are *water cooled*, as we would expect from Section 4.5.3.

It can be seen that the air delivery system (6) forms a substantial part of the system. At maximum power, we can speculate that the average cell voltage V_c will be about 0.62

volts. (We go a little above 0.6 as the system is pressurised, and fairly warm for a PEMFC.) The air stoichiometry λ will be about 2. Using equation A2.4 we can estimate the air flowrate to be:-

$$Air\ flowrate = 3.57\times 10^{-7}\times 2\ \times\frac{260000}{0.62}=0.3\ \text{kg.s}^{-1}$$

This figure is also given in the product information. In Chapter 8 we have derived equations for the power needed to compress air, and also the temperature rise. A reasonable estimate for the compressor efficiency η_c is 0.6. If we also assume that the entry temperature of the air is 300 K, and the pressure is 100 kPa, then, from equation 8.4 we have:-

$$\Delta T = \frac{300}{0.6}\left(\left(\frac{307}{100}\right)^{0.286}-1\right)= 189\ \text{K}$$

And also, from equation 8.7,

$$Compressor\ power = 1004\times\frac{300}{0.6}\left(\left(\frac{307}{100}\right)^{0.286}-1\right)\times 0.3 = 57000\ \text{Watts}$$

So we have a very substantial temperature rise, which will need to be compensated for by cooling the air before entry to the fuel cell. No doubt some of this cooling is accomplished by the humidification process. However, we still need 57 kW of power for the compressor. This is done in two stages. The first is a Lysholm or screw compressor, as in Section 8.2, and Figures 8.2(b) and 8.7, with the final pressure boosting being provided by a turbocharger, as in Sections 8.8 and 8.9. To find the power provided by the turbine we make the following assumptions:-

- The turbine efficiency is 0.6
- The exit temperature of the oxygen depleted air is 90 ^{0}C, 363K, which is the operating temperature of the fuel cell stack.
- The exit pressure is somewhat less than the entry pressure, otherwise there would be no through flow. Say 180 kPa above atmospheric, 280 kPa total.

The exit air flowrate will be greater than the entry air because of the addition of water, both because of humidification, and by the addition of product water. The mass flow rate is calculated as follows.

The exit air flowrate (excluding water) is given by equation A2.5, and using the same values as above, this gives:-

$$Exit\ air\ flowrate =(3.57\times10^{-7}\times2\ - 8.29\times10^{-8})\times\frac{260000}{0.62}=0.265\ \text{kg.s}^{-1}$$

The exit mass flow is increased by the water content. If we assume that the exit air is at about 100% humidity, so $P_w = 70$ kPa, from Table 4.1.

Since the total pressure of the exit air is 280 kPa, then the pressure of the dry air is thus $280 - 70 = 210$ kPa. Thus, from equation 4.4:-

$$\text{The humidity ratio } \omega = 0.622 \times \frac{70}{210} = 0.21$$

Hence the mass flowrate of water leaving the fuel cell is:-

$$\dot{m}_W = \omega \dot{m}_a = 0.21 \times 0.265 = 0.0557 \text{ kg.s}^{-1}, \text{ from equation 4.1.}$$

As an aside, we see that we can use this result to estimate the rate at which water was added to the reactant gases at entry. The rate of water production from the electrochemical reaction is given by equation A2.10 in Appendix 2, which gives, in this case:-

$$\textit{Rate of water production} = 9.34 \times 10^{-8} \times \frac{260000}{0.62} = 0.0392 \text{ kg.s}^{-1}$$

The difference between these two figures, namely $0.0557 - 0.0392 = 0.0165$ kg.s^{-1}, is an estimate of the rate at which water must enter the cell. Some of this water will have been in the air anyway, the rest is added by the humidification process. If the humidity of the entry air is about 70%, then it can be shown that approximately 2/3 of the water is added via humidification.[7] This example illustrates the beneficial effect of pressure on the humidification process. The proportion of excess water that would have been added if the pressure was lower would be much greater[8].

Returning to the exit air, the total exit flowrate is the dry air flowrate plus the water flowrate, giving $0.265 + 0.056 = 0.32$ kg.s^{-1}. This is only a little greater than the entry flowrate.

We can now use equation 8.10 to find the available turbine power. If we assume that the turbine exit pressure is still somewhat above air pressure, say 150 kPa, then:-

$$\textit{Turbine power} = 1004 \times 0.6 \times 363 \times \left(\left(\frac{280}{150} \right)^{0.286} - 1 \right) \times 0.32 = 14,000 \text{ Watts}$$

This 14 kW of power would make a useful contribution to the 57 kW needed. However, if we look at the temperature change, we see that it might well not be possible to

---

[7] It is left as an exercise for the reader to show this. All the necessary formulas are given in Section 4.4.2.

[8] Another useful exercise for the reader would be to repeat this calculation substituting some lower value for the exit total pressure, such as 120 kPa, for the 280 kPa used above. It will be found that the entry water rate will have to be about 0.19 kg.s^{-1}, that is greater by a factor of over 10.

harness all this power. From equation 8.9 we see that the temperature change through the turbine would be:-

$$\Delta T = 0.6 \times 363 \times \left(\left(\frac{280}{150} \right)^{0.286} - 1 \right) = 43 \ \text{K}$$

This would bring the exit gas temperature down to about 50 $^{\circ}$C. Bearing in mind that the air will have been more or less saturated as it left the fuel cell, we would anticipate a good deal of condensation in the turbine, which would inhibit its performance. We should therefore perhaps round down our estimated power from the turbine to the still by no means negligible 10 kW. The power from the motor driving the screw compressor will therefore be about 47 kW. This is a very substantial parasitic power loss, and largely explains why the traction motor mentioned above is rated at 160 kW, whereas the fuel cell is 260 kW. The other major loses are the cooling system (estimated at 20 kW parasitic loses) and the electrical sub-systems, estimated at 13 kW.

The fuel feed system (7) is one of the simplest parts of this fuel cell system. *Hydrogen compressed to a pressure of 24800 kPa is stored in cylinders on the roof of the bus. This is regulated down to the same pressure as the air feed, 207 kPa maximum, in two stages. The pressure regulation system incorporates an ejector,* as described in Section 8.10, *to circulate the hydrogen fuel through the stack.*

We might suppose that there would be a considerable temperature drop as the hydrogen gas pressure is reduced, which is what is normally observed with depressurising gases. If the hydrogen behaved as a perfect gas, then we could use the standard equation:-

$$\frac{T_2}{T_1} = \left(\frac{P_2}{P_1} \right)^{\frac{\gamma-1}{\gamma}}$$

for adiabatic changes to find the exit temperature – and we would estimate it to be about 80 K – very cold! In fact the hydrogen behaviour is far from that of a perfect gas. The so-called 'Joule-Thompson' effect comes in, and there is actually a very modest temperature rise, of about 7 $^{\circ}$C, in the pressure regulation system.

9.7   Closing Remarks

It is hoped that the description of a well engineered system given in the previous section has had the reader looking back and forth through many different parts of this book. It takes us back to where we started, illustrating that fuel cell systems require a broad engineering knowledge, and are highly interdisciplinary. We hope that we have introduced the many different technologies that underpin the operation of fuel cell systems. In all areas there is, of course, much more that could have been said, but having outlined the basic explanation of how and why things behave in the way they do, we hope that you have a good broad understanding of all the main components of a fuel cell system. If you find yourself specialising in one aspect of fuel cell technology, we hope that this book has helped you understand the work of your colleagues, and given you a good starting point for further studies in your particular area.

We are now witnessing the birth of a major new industry, one that has arisen through the dedicated effort of many people. Over the past few years there has been an unprecedented growth in interest, with major financial investments led by companies such as the vehicle makers DaimlerChrysler, Ford, General Motors and Toyota together with oil majors such as Shell, Atlantic Richfield (ARCO) and Texaco. As Geoffrey Ballard, founder of Ballard Power Systems has remarked, "the fuel cell did not come into being because one person had an idea." The dawn of the fuel cell industry has arisen "because dozens of people chose to work together to make an idea come into existence." (Koppel, 1999)

References

Auinger H., (1999) "Determination and designation of the efficiency of electrical machines" *Power Engineering Journal*, Vol .13, no. 1, pp15 – 23

Gibbard H.K. (1999) "Highly reliable 50 Watt fuel cell system for variable message signs", Proceedings of the European Fuel Cell Forum Portable Fuel Cells conference, Lucerne, pp 107-112

Harri V., Erni P., Egger S., (1999) "Super capacitors and batteries in applications with fuel cells", Proceedings of the European Fuel Cell Forum Portable Fuel Cells conference, Lucerne, pp 245 – 252

Hockaday R., Navas C., (1999) "Micro-fuel cells for portable electronics", Proceedings of the European Fuel Cell Forum Portable Fuel Cells conference, Lucerne, pp 45 - 54

Kenjo T. (1991) *Electric Motors and their Controls*, Oxford University Press

Kenjo T. (1994) *Power Electronics for the Microprocessor Age*, Oxford University Press

Koppel T. (1999) *Powering the future - the Ballard fuel cell and the race to change the world*, John Wiley & Sons, Canada Ltd.

Spiegel R.J., Gilchrist T., House D.E. (1999), "Fuel cell bus operation at high altitude", *Proceedings of the Institution of Mechanical Engineers*, Vol 213. part A, pp 57-68

Walters D. (1999) "Energy efficient motors, saving money or costing the earth?", *Power Engineering Journal*, Part 1 Vol 13 no.1, pp25 - 30, Part 2 Vol.13, no.2, pp44 - 48

Appendix 1

Change in Molar Gibbs Free Energy Calculations

A1.1   Hydrogen Fuel Cell

This section explains the calculations of $\Delta \bar{g}_f$ for the reaction

$$H_2 + \tfrac{1}{2}O_2 \rightarrow H_2O$$

which were given in Chapter 2.

The Gibbs function of a system is defined in terms of the entropy and the enthalpy:

$$G = H - TS$$

Similarly, the molar Gibbs energy of formation, the molar enthalpy of formation, and the molar entropy are connected by the equation:

$$\bar{g}_f = \bar{h}_f - T\bar{s}$$

In this case it is the **change** in energy that is important. Also, in a fuel cell, the temperature is constant. So we can say that:-

$$\Delta \bar{g}_f = \Delta \bar{h}_f - T\Delta \bar{s} \qquad \text{[A1.1]}$$

The value of $\Delta \bar{h}_f$ is the difference between $\bar{h}_f$ of the products and $\bar{h}_f$ for the reactants. Thus, for the reaction $H_2 + \tfrac{1}{2}O_2 \rightarrow H_2O$ we have:-

$$\Delta \bar{h}_f = \left(\bar{h}_f\right)_{H_2O} - \left(\bar{h}_f\right)_{H_2} - \tfrac{1}{2}\left(\bar{h}_f\right)_{O_2} \qquad \text{[A1.2]}$$

Similarly, $\Delta \bar{s}$ is the difference between $\bar{s}$ of the products and $\bar{s}$ of the reactants. So, in this case:-

$$\Delta \bar{s} = (\bar{s})_{H_2O} - (\bar{s})_{H_2} - \tfrac{1}{2}(\bar{s})_{O_2} \qquad \text{[A1.3]}$$

The values of $\bar{h}_f$ and $\bar{s}$ vary with temperature according to the equations given below. These standard equations are derived using thermodynamic theory, and their proof can be found in most books on engineering thermodynamics (e.g. Balmer, 1990). In these equations the subscript to $\bar{h}$ and $\bar{s}$ is the temperature, and $\bar{c}_p$ is the molar heat capacity at constant pressure. 298.15 K is standard temperature.

The molar enthalpy of formation at temperature T is given by:-

$$\bar{h}_T = \bar{h}_{298.15} + \int_{298.15}^{T} \bar{c}_p dT \qquad \text{[A1.4]}$$

The molar entropy is given by:-

$$\bar{s}_T = \bar{s}_{298.15} + \int_{298.15}^{T} \frac{1}{T}\bar{c}_p dT \qquad \text{[A1.5]}$$

The values for the molar entropy and enthalpy of formation at 298.15K are obtainable from thermodynamics tables (e.g. Keenan and Kaye, 1948), and are given in Table A1 below. These values are at standard pressure.

Table A1.1 Values of $\bar{h}_f$ in J.mol^{-1} and $\bar{s}$ in J.mol^{-1}.K^{-1}, at 298.15 K, for the hydrogen fuel cell

	h_f	$\bar{s}$
H_2O (liquid)	-285838	70.05
H_2O (steam)	-241827	188.83
H_2	0	130.59
O_2	0	205.14

To use equations A1.4 and A1.5 we need to know the values of the molar heat capacity at constant pressure $\bar{c}_p$. Over a range of temperatures $\bar{c}_p$ is not constant. Nevertheless, empirical equations for $\bar{c}_p$ are obtainable and given in many thermodynamics texts (e.g. Van Wylen, 1986), and the equations given below are accurate to within 0.6% over the range 300 to 3500K.

For steam:-

$$\bar{c}_p = 143.05 - 58.040T^{.25} + 8.2751T^{.5} - 0.036989T$$

For hydrogen, H_2 :-

$$\bar{c}_p = 56.505 - 22222.6\,T^{-.75} + 116500\,T^{-1} - 560700\,T^{-1.5}$$

For oxygen, O_2 :-

$$\bar{c}_p = 37.432 + 2.0102 \times 10^{-5}\,T^{1.5} - 178570\,T^{-1.5} + 2368800\,T^{-2}$$

All these equations for $\bar{c}_p$ are in J/gmole K. They can be substituted into equations A1.4 and A1.5, giving functions that can be readily integrated and thus evaluated at any temperature T. This is done to derive values for $\bar{h}_f$ and $\bar{s}$ for steam, hydrogen and oxygen. These values are then substituted into equations A1.2 and A1.3. This gives values for $\Delta\bar{h}_f$ and $\Delta\bar{s}$, which are finally substituted into equation A1.1, giving us the change in molar Gibbs energy of formation $\Delta\bar{g}_f$. Sample values are shown in Table A1.2 below.

For the case of liquid water, standard values from Table A.1 for $\bar{h}_f$ and $\bar{s}$ are used for 25 °C. At 80 °C equations A1.4 and A1.5 are again used to find $\bar{h}_f$ and $\bar{s}$, but in this case we can assume that $\bar{c}_p$ is constant, since we are working over such a small temperature range.

Table A1.2 Sample values for $\Delta\bar{h}_f$, $\Delta\bar{s}$, and $\Delta\bar{g}_f$, for the reaction $H_2 + \frac{1}{2}O_2 \rightarrow H_2O$. Temperatures are in Celsius, other figures are in kJ.mol⁻¹.

Temperature	$\Delta\bar{h}_f$	$\Delta\bar{s}$	$\Delta\bar{g}_f$
100	- 242.6	- 0.0466	- 225.2
300	- 244.5	- 0.0507	- 215.4
500	- 246.2	- 0.0533	- 205.0
700	- 247.6	- 0.0549	- 194.2
900	- 248.8	- 0.0561	- 183.1

A1.2   The Carbon Monoxide Fuel Cell

It is possible that in the higher temperature fuel cells introduced in chapter 6 the carbon monoxide gas generated from steam reforming of fuel such as methane is directly oxidised. The reaction is:-

$$CO + \tfrac{1}{2}O_2 \rightarrow CO_2$$

The method used, and the theory employed, for calculating the Gibbs free energy change is exactly the same as for the hydrogen fuel cell, except that the equations are altered to fit the new reaction. Oxygen features again as it did before, and the values of the molar specific heat capacity for carbon monoxide and carbon dioxide are given by:-

For CO:- $\bar{c}_p = 69.145 - 0.022282\,T^{.75} - 2007.7\,T^{-0.5} + 5589.64\,T^{-0.75}$

For CO_2:- $\bar{c}_p = -3.7357 + 3.0529\,T^{.5} - 0.041034\,T + 2.4198\times10^{-6}\,T^{2}$

Together with values from Table A1.3 below, these equations are used with equations A1.4 and A1.5 to find the molar enthalpies and entropies for the three gases in question.

Table A1.3 Values of $\bar{h}_f$ in J.mol^{-1} and $\bar{s}$ in J.mol^{-1}.K^{-1}, at 298.15 K, for the carbon monoxide fuel cell.

	h_f	$\bar{s}$
O_2	0	205.14.
CO	-110529	197.65
CO_2	-393522	213.80

The change in the molar enthalpy and molar entropy is then calculated using these two equations:-

$$\Delta\bar{h}_f = \left(\bar{h}_f\right)_{CO_2} - \left(\bar{h}_f\right)_{CO} - \tfrac{1}{2}\left(\bar{h}_f\right)_{O_2}$$

$$\Delta\bar{s} = \left(\bar{s}\right)_{CO_2} - \left(\bar{s}\right)_{CO} - \tfrac{1}{2}\left(\bar{s}\right)_{O_2}$$

The change in molar Gibbs free energy of formation is then calculated, as with the hydrogen fuel cell, using equation A1.1 Some example results are given below in Table A1.4

Table A1.4 Sample values for $\Delta\bar{h}_f$, $\Delta\bar{s}$, and $\Delta\bar{g}_f$, for the reaction $CO + \tfrac{1}{2}O_2 \rightarrow CO_2$. Temperatures are in Celsius, other figures are in kJ.mol^{-1}.

Temperature	$\Delta\bar{h}_f$	$\Delta\bar{s}$	$\Delta\bar{g}_f$
100	- 283.4	- 0.0877	- 250.7
300	- 283.7	- 0.0888	- 232.7
500	- 283.4	- 0.0890	- 214.6
700	- 282.8	- 0.0887	- 196.5
900	- 282.0	- 0.0883	- 178.5

References

Balmer R., (1990), *Thermodynamics*, West, Chapter 6

Keenan J.H & Kaye J., (1948) *Gas Tables*, Wiley,

Van Wylen G.J.& Sonntag R.E., (1986), *Fundamentals of Classical Thermodynamics*, 3rd Ed., Wiley, p.688

Appendix 2

Useful Fuel Cell Equations

A2.1   Introduction

In this appendix many useful equations are derived. They relate to:-

- Oxygen usage rate
- Air inlet flowrate
- Air exit flowrate
- Hydrogen usage, and the energy content of hydrogen
- Rate of water production
- Heat production

In many of the sections that follow term *stoichiometric* is used. Its meaning could be defined as "just the right amount". So, for example, in the simple fuel cell reaction:

$$2\,H_2 + O_2 \rightarrow 2\,H_2O$$

exactly 2 moles of hydrogen would be provided for each mole of oxygen. This would produce exactly 4 Faradays of charge, since two electrons are transferred for each mole of hydrogen. Note that either or both the hydrogen and oxygen are often supplied at greater than the stoichiometric rate. This is especially so for oxygen if it is being supplied as air. If it were supplied at exactly the stoichoimetric rate then the air leaving the cell would be completely devoid of oxygen. Note also that reactants cannot be supplied at *less* than the stoichiometric rate.

To increase the usefulness of the formulas, they have been given in terms of the electrical power of the whole fuel cell stack P_e, and the average voltage of each cell in the stack V_c. The electrical power will nearly always be known, as it is the most basic and important information about a fuel cell system. If V_c is not given it can be assumed to be between 0.6 and 0.7 volts, as most fuel cells operate in this region (see Figures 3.1 and 3.2). If the efficiency is given, then V_c can be calculated using equation 2.5. If no figures are given then using $V_c = 0.65$ volts will give a good approximation. Estimate somewhat higher if the fuel cell is pressurised.

A2.2   Oxygen and Air Usage

From the basic operation of the fuel cell, we know that

$$charge = 4\,F \times amount\ of\ O_2$$

Dividing by time, and rearranging

$$O_2\ usage = \frac{I}{4F}\ \ moles.s^{-1}$$

This is for a single cell. For a stack of n cells:-

$$O_2\ usage = \frac{I\,n}{4F}\ \ moles.s^{-1} \qquad\qquad [A2.1]$$

However, it would be more useful to have the formula in kg.s^{-1}, without needing to know the number of cells, and in terms of power, rather than current. If the voltage of each cell in the stack is V_c , then:

$$Power,\ P_e = V_c \times I \times n$$

So,
$$I = \frac{P_e}{V_c \times n}$$

Substituting this into equation A2.1 gives :

$$O_2\ usage = \frac{P_e}{4.V_c.F}\ \ moles.s^{-1} \qquad\qquad [A2.2]$$

Changing from moles.s^{-1} to kg.s^{-1} :-

$$O_2\ usage = \frac{32 \times 10^{-3}.P_e}{4\ V_c\ F}\ \ kg.s^{-1}$$

$$= 8.29 \times 10^{-8} \times \frac{P_e}{V_c}\ kg.s^{-1} \qquad\qquad [A2.3]$$

This formula allows the oxygen usage of any fuel cell system of given power to be calculated. If V_c is not given, it can be calculated from the efficiency, and if that is not given, the figure of 0.65 volts can be used for a good approximation.

However, the oxygen used will normally be derived from air, so we need to adapt equation A2.2 to air usage. The molar proportion of air that is oxygen is 0.21, and the molar mass of air is 28.97×10^{-3} kg.mole^{-1}. So, equation A2.2 becomes:

$$Air\ usage = \frac{28.97 \times 10^{-3} \times P_e}{0.21 \times 4 \times V_c \times F}\ \text{kg.s}^{-1}$$

$$= 3.57 \times 10^{-7} \times \frac{P_e}{V_c}\ \text{kg.s}^{-1}$$

However, if the air was used at this rate, then as it left the cell it would be completely devoid of any oxygen - it would all have been used. This is impractical, and in practice the air flow is well above stoichiometric, typically twice as much. If the stoichiometry is λ, then the equation for air usage becomes:-

$$Air\ usage = 3.57 \times 10^{-7} \times \lambda \times \frac{P_e}{V_c}\ \text{kg.s}^{-1} \qquad \text{[A2.4]}$$

The kilogram per second is not, in fact, a very commonly used unit of mass flow. The following conversions to "volume at standard conditions related" mass flow units will be found useful. The mass flowrate from equation A2.4 should be multiplied by:-

- 3050 to give flowrate in standard m^3.hr^{-1}
- 1795 to give flowrate in SCFM (or standard ft^3.min^{-1})
- 5.1×10^4 to give flowrate in slm (standard L.min^{-1})
- 847 to give flowrate in sls (standard L.sec^{-1})

A2.3   Air Exit Flowrate

It is sometimes important to distinguish between the *inlet* flowrate of the air, which is given by equation A2.4 above, and the *outlet* flowrate. This is particularly important when calculating the humidity, which is an important issue in certain types of fuel cell, especially PEM fuel cells. The difference is caused by the consumption of oxygen. There will usually be more water vapour in the exit air, but we are considering "dry air" at this stage. Water production is given in section A2.5 below. Clearly:-

Exit air flowrate = Air inlet flowrate − oxygen usage

Using equations A2.3 and A2.4 this becomes:-

$$Exit\ air\ flowrate = 3.57 \times 10^{-7} \times \lambda \times \frac{P_e}{V_c} - 8.29 \times 10^{-8} \times \frac{P_e}{V_c}\ \text{kg.s}^{-1}$$

$$= \left(3.57 \times 10^{-7} \times \lambda - 8.29 \times 10^{-8}\right) \times \frac{P_e}{V_c}\ \text{kg.s}^{-1} \qquad \text{[A2.5]}$$

A2.4   Hydrogen Usage

The rate of usage of hydrogen is derived in a similar way to oxygen, except that there are two electrons from each mole of hydrogen. Equations A2.1 and A2.2 thus become:

$$H_2 \ usage = \frac{I\,n}{2\,F} \ \text{moles.s}^{-1} \qquad\qquad [A2.6]$$

and

$$H_2 \ usage = \frac{P_e}{2\,V_c\,F} \ \text{moles.s}^{-1} \qquad\qquad [A2.7]$$

The molar mass of hydrogen is 2.02×10^{-3} kg.mole^{-1}, so this becomes:

$$H_2 \ usage = \frac{2.02 \times 10^{-3}.P_e}{2\,V_c\,F}$$

$$= 1.05 \times 10^{-8} \times \frac{P_e}{V_c} \ \text{kg.s}^{-1} \qquad\qquad [A2.8]$$

at stoichiometric operation. Obviously, this formula only applies to a hydrogen-fed fuel cell. In the case of a hydrogen/carbon monoxide mixture derived from a reformed hydrocarbon, things will be different, depending on the proportion of carbon monoxide present. The result can be transformed to a volume rate using the density of hydrogen, which is 0.084 kg.m^{-3} at NTP.

As well as the rate of usage of hydrogen, it is often also useful to know the electrical energy that could be produced from a given mass or volume of hydrogen. The list below gives the energy in kWh, rather than Joules, as this in the more usual measure used for electrical power systems. As well as the "raw" energy per kilogram and standard litre, we have also given an "effective" energy, taking into account the efficiency of the cell. This is given in terms of V_c, the mean voltage of each cell. If an equation with the efficiency is needed, then use the formula derived in Section 2.5, page 27,

$$efficiency = \frac{V_c}{1.48}$$

Table A2.1 "Raw" and effective energy content of hydrogen fuel

Form	Energy content
Specific enthalpy (HHV)	1.42 J.kg^{-1}
Specific enthalpy (HHV)	39.3 kWh. kg^{-1}
Effective specific electrical energy	$26.6 \times V_c$ kWh.kg^{-1}
Energy density at STP (HHV)	3.20 kWh.m^{-3} = 3.20 Wh.SL^{-1}
Energy density at NTP (HHV)	3.29 kWh.m^{-3} = 3.29 Wh.SL^{-1}

A2.5   Water Production

In a hydrogen-fed fuel cell water is produced at the rate of one mole for every two electrons. (Revisit Section 1.1 if you are not clear why.) So, we again adapt equation A2.2 to obtain:

$$Water \ production = \frac{P_e}{2. \ V_c. \ F} \ moles.s^{-1} \qquad [A2.9]$$

The molecular mass of water is 18.02×10^{-3} kg.mole^{-1}, so this becomes:

$$Water \ production = 9.34 \times 10^{-8} \times \frac{P_e}{V_c} \ kg.s^{-1} \qquad [A2.10]$$

In the hydrogen-fed fuel cell the rate of water production more or less has to be stoichiometric. (One exception, which would raise this figure very slightly, is if the "self-humidifying" electrolyte discussed towards the end of the Section 4.4.4 were used.) However, if the fuel cell is a mixture of carbon monoxide with hydrogen, then the water production would be less – in proportion to the amount of carbon monoxide present in the mixture. If the fuel were a hydrocarbon internally reformed then the some of the product water would be used in the reformation process. We saw in Chapter 6, for example, that if methane is internally reformed, then half the product water in used in the reformation process, thus halving the rate of production.

It is sometimes useful to give an example figure to clarify a formula such as this. Let us take as an example a 1 kW fuel cell operating for 1 hour, at a cell voltage of 0.7 volts. This corresponds to an efficiency of 47% ref. HHV (from equation 2.5). Substituting this into equation A2.9 gives:-

$$The \ rate \ of \ water \ production = 9.34 \times 10^{-8} \times \frac{1000}{0.7}$$
$$= 1.33 \times 10^{-4} \ kg.s^{-1}$$

So the mass of water produced in one hour is:

$$= 1.33 \times 10^{-4} \times 60 \times 60 = 0.48 \ kg$$

Since the density of water is 1.0 g.cm^{-3}, this corresponds to 480 cm^3, which is almost exactly 1 pint. So, as a rough guide, 1 kWh of fuel cell generated electricity produces about 1 pint or 0.5 litres of water.

A2.6   Heat Produced

Heat is produced when a fuel cell operates. It was noted in Chapter 2, section 2.4, that if all the enthalpy of reaction of a hydrogen fuel cell were converted into electrical energy then the output voltage would be:-

> 1.48 volts if the water product were in liquid form
> or   1.25 volts if the water product were in vapour form.

It clearly follows that the difference between the actual cell voltage and this voltage represents that energy that is not converted into electricity - i.e. the energy that is converted into heat instead.

The cases where water finally ends in liquid form are so few and far between as to be not worth considering. So we will restrict ourselves to the vapour case. However, please note that this means we have taken into account the cooling effect of water evaporation. It also means that energy is leaving the fuel cell in three forms, electricity, ordinary "sensible" heat, and in the latent heat of the water vapour.

For a stack of n cells at current I, the heat generated is thus:-

$$Heating \quad rate = n I \, (1.25 - V_c) \quad Watts$$

In terms of electrical power, this becomes:-

$$Heating \quad rate = P_e \left(\frac{1.25}{V_c} - 1 \right) \quad Watts \qquad [A2.11]$$

Index